D1283050

Lecture Notes in Economics and Mathematical Systems
Operations Research, Computer Science, Social Science

Edited by M. Beckmann, Providence, G. Goos, Karlsruhe, and H. P. Künzi, Zürich

84

A. V. Balakrishnan

# Stochastic Differential Systems I

## Filtering and Control
## A Function Space Approach

Springer-Verlag
Berlin · Heidelberg · New York 1973

Dr. A.V. Balakrishnan
System Science Department
School of Engineering and Applied Sciences
University of California
Los Angeles, Calif. 90024/USA



AMS Subject Classifications (1970): 60G05, 60G35, 60G45, 49E99

ISBN 3-540-06303-X Springer-Verlag Berlin · Heidelberg · New York
ISBN 0-387-06303-X Springer-Verlag New York · Heidelberg · Berlin



Offsetdruck: Julius Beltz, Hemsbach/Bergstr.

# PREFACE

This book is an outgrowth of a graduate course by the same title given at UCLA (System Science Department), presenting a Functional Analysis approach to Stochastic Filtering and Control Problems. As the writing progressed, several new points of view were developed and as a result the present work is more in the nature of a monograph on the subject than a distilled compendium of extant works.

The subject of this volume is at the heart of the most used part of modern Control Theory - indeed, the bread-and-butter part. It includes the Linear (Bucy-Kalman) Filter Theory, the Feedback Control (regulation and tracking) Theory for plants with random disturbances, and Stochastic Differential Games. Linear Filter Theory is developed by a 3-Martingale approach and is perhaps the sleekest one to date. We hasten to add that although the terms are Engineering-oriented, and a background in Control Engineering is essential to understand the motivation, the work is totally mathematical, and in fact our aim is a rigorous mathematical presentation that is at once systematic.

We begin with some preliminary necessary notions relating to Stochastic Processes. We follow Parthasarathy's work in inducing Wiener measure on the Banach Space of Continuous functions. We introduce the linear Stochastic integrals right away. We are then ready to treat linear Stochastic Differential Equations. We then look at the measures induced, and in particular the Radon-Nikodym derivatives with respect to Wiener measure. This leads us naturally to consider Ito integrals, one advantage of which is shown in the simpler form of Radon-Nikodym derivatives obtainable in this way. We obtain R-N derivatives for Volterra-type operators. The Kalman Filtering Theory is considered next as an example and as an alternate means of getting the R-N derivative for Differential Systems. We study asymptotic properties of stable time-invariant systems, indicating at the same time an iterative technique for solving the Riccati Equation that occurs, including the asymptotic case. We next consider a variety of Stochastic Control problems. Here again we present a fresh, systematic approach. The main feature here is that we make no use of the dynamic programming formalism, unlike most current treatments. Moreover in our development, the so-called 'separation principle' is an easy byproduct. We also present a theory of stochastic Differential Games with imperfect observation. The final Chapter deals with the problem of Identifying a Linear Dynamic System from external measurements. In the Supplementary Notes the discrete versions of recursive filtering and likelihood ratios are treated, leading to an approximation of Ito integrals by discrete sums necessary in digital computation.

We do not consider Fokker-Planck equations since we feel they are more useful in the non-linear case and thus more properly belong in the second volume of the book devoted to the non-linear equations.

It is the author's pleasant duty to acknowledge the many stimulating discussions with Professor R. E. Mortensen and Jiri Ruzicka and to thank Trudy Cook for typing the several versions of the Notes with patience and forebearance.

Los Angeles, November, 1971

CONTENTS

# CHAPTER I

## PRELIMINARIES: STOCHASTIC PROCESSES

We begin with some preliminary review material on stochastic processes, fixing some of the vocabulary at the same time. The standard reference is Doob [2], supplemented by Gikhman-Skorokhod [3] where sometimes an updated presentation is helpful. Naturally we shall only touch upon aspects of direct concern to us.

A stochastic process - in the axiomatic approach - is an indexed family of random variables. The index or parameter set will, for us, be an interval of real numbers, perhaps infinite. It will always indicate time, and hence the parameter set will be denoted $T$. The range of the random variables will be a finite-dimensional Euclidean space, denoted $E$. In the customary notation, we have then a family of functions $f(t;\omega)$, $t$ denoting the parameter, and for each $t$ we have measurable function on a probability measure space $\Omega$, $\omega \in \Omega$, with a sigma-algebra of subsets $\mathscr{B}$ and probability measure $p(\cdot)$. We can of course artificially construct (physically plausible) such processes. For example, let $\theta$ denote a random variable uniformly distributed between zero and $2\pi$,

$$f(t;\theta) = \sin(2\pi t + \theta)$$

or,

$$f(t;\theta) = h(t + \theta)$$

where $h(\cdot)$ is a function defined on the real line to be one on the rationals and zero otherwise. The second example is merely to serve to illustrate a point below.

Practical stochastic processes do not of course all come this way. The axiomatic set-up then has to be manufactured. Thus we are usually given (or can deduce) 'consistent' joint distributions of a finite number of variables selected arbitrarily from the parameter set; that is all. In that case we proceed as follows: we look at the 'time-histories', or 'realizations' of the process as time evolves; which will simply be a class of time functions. Let $X$ denote the class of all functions (with range in E) defined on T. For each $t$, we consider the class of sets in X which are inverse images of Borel sets in E, and let $\mathcal{S}$ denote the smallest sigma-algebra containing all these sets as $t$ runs over T. Let $t_k$, $k = 1, \ldots n$ be a finite number of time points, and let $P_n(\cdot)$ denote the corresponding joint distribution. Let $I_k$, $k = 1, \ldots n$ denote intervals in E. Then we define a set function:

$$p[f \mid f(\cdot) \in X, f(t_k) \in I_k, k=1, \ldots n]$$

$$= \int_{I_1} \ldots\ldots \int_{I_n} dP_n \qquad (1.1)$$

The celebrated extension theorem of Kolmogorov [1] then tells us that $p(\cdot)$ can be extended to be a countably additive measure on $\mathcal{S}$. The student will do well to understand thoroughly the procedure involved, since it is basic to the axiomatic theory. Note that the measure is uniquely determined by the finite-dimensional distributions in the sense that any two extensions must agree on $\mathcal{S}$.

Note also that we have in this manner produced a probability measure space supporting a family of random variables whose finite-dimensional distributions coincide with the given distributions, by simply taking the 'function-value' at each $t$. We have thus produced the axiomatic set-up postulated. Since the 'sample-space' is now the function space $X$, such a process is referred to as a "Function Space" Process. If we had begun with the axiomatic set-up, then the mapping $\omega \rightarrow f(t;\omega)$ is a measurable mapping of $\Omega$ into X. It is interesting to see what happens to the second example if we followed the construction beginning with the finite dimensional distributions. Every cylinder set has measure one or zero depending on whether it contains zero function or not. Consider the class of sets in $\mathscr{S}$ with the property that either the set contains the zero function and has unit measure, or it does not contain the zero function and has zero measure. This class is clearly a sigma-algebra, and it contains all cylinder sets. Hence it is all of $\mathscr{S}$. In particular then any set in $\mathscr{S}$ that contains the zero function has unit measure by simply defining the measure of a set as one if it contains the zero function, and zero otherwise, we obtain a probability measure defined on all the subsets of X. On the other hand, $f(t;\theta)$ is not identically zero for any $\theta$! Suppose we define a new process $\tilde{f}(t;\theta)$ identically zero. For each t,

$$[\theta \mid f(t;\theta) \neq \tilde{f}(t;\theta)]$$

has zero probability, and hence

$$f(t;\theta) = \tilde{f}(t;\theta)$$

for each t, with probability one so that in particular the finite dimensional distributions corresponding to $f(t;\theta)$ and $\tilde{f}(t;\theta)$ agree. It is then natural to say that the process $f(t;\theta)$ is 'equivalent' to the process $\tilde{f}(t;\theta)$.

The process $\tilde{f}(t;\theta)$ is "separable' - that is to say, we can find a countable dense set of points $\{t_j\}$ in T and a set $N \in \mathscr{S}$ of measure zero such that for any open set G in T and an arbitrary closed set F in E, the set

$$[\theta \mid f(t;\theta) \in F \text{ for all } t \text{ in } G]$$

differs from the set

$$[\theta \mid f(t;\theta) \in F \quad \text{for all} \quad t_j \in G]$$

by a subset of N. The main point here is that Doob has shown that given any process [with range in a locally compact space as E] we can always construct an equivalent separable process. The advantage of a separable process is that probabilities of events depending on a non-countable number of time points can be obtained now as those depending only on a countable number, and this in turn facilitates operations. It should be noted however that in the construction of the equivalent separable process we may introduce 'sample paths' not in the original definition (as in our example above where the zero function was added). For details of the construction, see Doob [2] or Gikhman-Skorokhod [3], or Neveu [24].

Note that a process is separable if it is continuous with probability one - [$f(t;\omega)$ is continuous in $t$ for almost all $\omega$]. In this book we shall mainly be concerned with processes continuous with probability one. In such a case it is natural to try and confine oneself to a smaller space than X, and furthermore exploit any topology we can thereby obtain.

<u>Notation</u> For any m-by-p matrix $x$, we shall use the notation:

$$\|x\|^2 = \text{Tr. } x x^*$$

$*$ denoting 'transpose', Tr. denoting 'trace'

We shall deal with real variables only throughout. For any two m-by-p matrices $x, y$ we shall use the 'inner-product' notation:

$$[x, y] = \text{Tr}\, x y^* = \text{Tr}\, x^* y = \text{Tr}\, y x^*$$

We shall use E(.) to denote the 'expected value' of a random variable. A random variable will be said to have a finite first moment if

$$E(\|x\|) < \infty$$

An element in real Euclidean space of dimension n will be written as an n-by-one matrix. A random variable will be said to have a finite second moment if

$$E(\|x\|^2) < \infty$$

The following inequalities (named after Schwarz and Jensen respectively)

$$| E( [x, y] ) | \leq \sqrt{E(\|x\|^2)} \sqrt{E(\|y\|^2)}$$

$$\| E(x) \|^2 \leq E(\|x\|^2)$$

will often be used. For instance, because of these inequalities, we can define , for any two random variables with finite second moments, the 'covariance matrix':

$$E( x y^* ) - E(x) E(y)^*$$

Example Gaussian Processes

A stochastic process $x(t;\omega)$ is said to be Gaussian if the joint distribution of any finite number of variables $x(t_k;\omega)$ is Gaussian. We shall consider only processes with finite second moments: i.e.:

$$E( \|x(t;\omega)\|^2 ) < \infty \quad \text{for every } t \in T$$

Let

$$m(t) = E( x(t;\omega) )$$

The function m(t) is then referred to as the mean function. Similarly, the covariance function is :

$$E( (x(t;\omega) - m(t) ) (x(s;\omega) - m(s))^* ) , \qquad t, s \in T$$

and is of course also equal to:

$$E( (x(t;\omega) x(s;\omega)^* ) - m(t) m(s)^*$$

Let R(t;s) denote the covariance function. Then R(t;s) is a 'positive definite' function; that is, it has the properties:

(i) $R(t;s) = R(s;t)^*$

(ii) for any finite set of points $t_k$ , and arbitrary real constants $a_k$ ,

$\sum \sum a_i a_j R(t_i;t_j)$ is self-adjoint and non-negative definite.

The main point about the Gaussian processes is that they are completely determined by the mean and covariance. In fact given a positive definite matrix function $R(t; s)$, $t, s \in T$, with properties (i) and (ii); let $\mu(t)$ be any function on $T$ with range in E. Then there exists a Gaussian ('Function-Space', Separable) stochastic process $x(t;\omega)$ with range in E such that

$$E(x(s;\omega)x(t;\omega)^*) = R(s;t) + \mu(s)\,\mu(t)^*$$

$$E(x(s;\omega)) = \mu(s)$$

See Doob [2, p. 72]. The main point is that a Gaussian distribution is completely specified by the mean and variance.

Let $x(t;\omega)$ be a separable stochastic process with mean zero and covariance function $R(s:t)$ which we shall now assume to be (locally) continuous, and take $T$ to be an interval. Then by the Chebychev inequality, $x(t;\omega)$ is continuous in probability (or stochastically continuous), that is to say, for any $\epsilon > 0$

$$p(\|x(t;\omega)-x(t+\Delta;\omega)\| \geq \epsilon)$$

goes to zero with $|\Delta|$. Hence [Doob [2], p. 61] $x(t;\omega)$ is measurable jointly in $\omega$ and $t$ (with respect to Lebesgue measure). Let us next specialize to the case where $T$ is compact. We have then the Karhunen-Loeve expansion: (See e.g., [2, 3, 10]):

$$x(t;\omega) = \sum_{1}^{\infty} \zeta_n(\omega)\, \phi_n(t) \tag{1.2}$$

where the series converges in the 'mean square' sense:

$$E(\|x(t;\omega) - \sum_{1}^{m} \zeta_k(\omega)\, \phi_k(t)\|^2)$$

$$\longrightarrow 0 \;,$$

$$\zeta_n(\omega) = \int_T [\phi(t),\; x(t;\omega)]dt,$$

and $\phi_n(\cdot)$ is an orthonormal sequence of eigen functions of the ( non-negative def., compact) operator R, mapping $L_2(T)$ into $L_2(T)$, defined by:

$$Rf = g; \;\; g(t) = \int_T R(t;s)\, f(s)ds \qquad t \in T$$

The convergence follows from the readily verified fact that

$$E(\zeta_n(\omega)^2) = \lambda_n = \text{eigen-value correspond to } \phi_n(\cdot)$$

and

$$E(\|x(t;\omega) - \sum_{1}^{n} \zeta_k(\omega)\, \phi_k(t)\|^2) = R(t;t) - \sum_{1}^{n} \lambda_k[\phi_k(t), \phi_k(t)]$$

$$\longrightarrow 0 \text{ uniformly for } t \text{ in } T$$

If the process $x(t;\omega)$ is Gaussian in addition, the variables $\zeta_k(\omega)$ are independent (being uncorrelated Gaussians), and the series then converges also with probability one by virtue of the Kolmogorov inequality, which because of its subsequent use, we shall now state as a theorem:

<u>Theorem 1.1</u> (Kolmogorov Inequality) Let $y_i$, $i=1, \ldots n$ be n

independent zero mean variables with range in E, and variance $\sigma_i^2$. Let

$$x_k = \sum_1^k y_i$$

Then

$$p(\max_{k=1,\dots n} \|x_k\| \geq \epsilon) \leq (1/\epsilon^2) \sum_1^n \sigma_k^2$$

Proof  See e.g., Loeve [10], page 235.

Let $A_k = [\|x_k\| \geq \epsilon]$

Let $B_k$ denote the set where only $\|x_k\|$ is $\geq \epsilon$:

$$B_k = [\|x_1\| < \epsilon, .. \|x_{k-1}\| < \epsilon, \|x_k\| \geq \epsilon]$$

Then

$$B_k = A_k - \bigcup_1^{k-1} A_j$$

and

$$E[\|x_n\|^2] \geq \sum_1^n \int_{B_k} \|x_n\|^2 dp = \sum_1^n \int_{B_k} \left\{ \|x_n - x_k\|^2 + \|x_k\|^2 + 2[x_k, x_n - x_k] \right\} dp .$$

Now define

$$z_k = x_k \text{ on } B_k$$

$$= 0 \text{ otherwise}$$

Then

$$\int_{B_k} [x_k, x_n - x_k] dp = \int_\Omega [z_k, (x_n - x_k)] dp = 0$$

since $z_k$ is independent of $(x_n - x_k) = \sum_{k+1}^n y_j$

Hence

$$\sum_1^n \sigma_i^2 = E[\|x_n\|^2] \geq \sum_1^n \int_{B_k} \|^2 dp \geq \varepsilon^2 \, p[\cup_k B_k] = \varepsilon^2 \, p[\cup_k A_k]$$

$$= \varepsilon^2 \, p[\max \|x_k\| \geq \varepsilon]$$

which yields the required result.

Corollary Let $y_i$ be zero mean independent random variables such that

$$\sum_1^\infty \sigma_i^2 < \infty, \qquad \sigma_i^2 = E[\|y_i\|^2]$$

Then

$$x_n = \sum_1^n y_i \text{ converges with probability one}$$

Proof The set of non-convergence

$$= \bigcup_k \bigcap_n \bigcup_\nu \left[\|x_{n+\nu} - x_n\| \geq \frac{1}{k}\right]$$

Now

$$p\left(\bigcup_\nu \|x_{n+\nu} - x_n\| \geq \frac{1}{k}\right)$$

$$= \lim_{N \to \infty} p\left[\bigcup_{\nu=1}^N \|x_{n+\nu} - x_n\| \geq \frac{1}{k}\right]$$

$$\leq \lim_{N \to \infty} k^2 \sum_{n+1}^N \sigma_i^2 = k^2 \sum_{n+1}^\infty \sigma_i^2$$

using Kolmogorov inequality, and hence etc.

Example Let $W(t;\omega)$ be a Gaussian process with mean zero and covariance:

$$R(s;t) = \min(s;t)I, \qquad s, t \geq 0$$

where $I$ is the identity matrix and $T$ is the interval $[0, 1]$. The corresponding eigenfunctions are well-known. Taking the dimension of

E to be n, we have

$$\phi_k(t) = \text{col } \varphi_{k,i}(t), \quad i = 1, \ldots n$$

where

$$\varphi_{k,i}(t) = \sqrt{\frac{2}{n}} \sin (2k+1) \pi t/2$$

$$E[\zeta_k^2] = 4/(2k+1)^2 \pi^2$$

so that

$$W(t;\omega) = \sum_1^{\infty} \zeta_k(\omega) \phi_k(t) \tag{1.3}$$

where the series converges in the mean square and with probability one.

# CHAPTER II

## LINEAR STOCHASTIC EQUATIONS

In this chapter we shall show how under a sufficient condition a stochastic process induces a probability measure on the sigma-algebra of Borel sets of the Banach space of continuous functions. A special case of central importance is the Wiener measure, or equivalently, the Wiener Process. We can then define stochastic integrals with respect to the Wiener process, and in turn this leads us to linear stochastic differential systems with the Wiener process as the "forcing term."

### 2.1 Inducing Measures on C

Given a stochastic process, or equivalently, a consistent family of finite dimensional distributions, we have seen that we can always construct a "function space" process with $X$ as the sample space and a probability measure $p(.)$ on the sigma-algebra $'\mathscr{S}$ (that "agrees" with the finite dimensional distributions). This space is too large for our purposes; in our study of stochastic equations we shall be concerned only with processes for which we can confine the sample space to the class of continuous functions. A sufficient condition that insures this is the following: for any two arbitrary time-points $t_1$, $t_2$, denoting the corresponding "variables" by $x(t_1)$, $x(t_2)$, we have:

$$E(|x(t_2) - x(t_1)|^r) \leq k \; |t_2 - t_1|^{1+\delta} \tag{2.1}$$

where

$r > 0, k > 0, \delta > 0$ are fixed constants independent of $t_1, t_2$; and $|\cdot|$ denotes the Euclidean norm.

Furthermore we shall assume that T is a compact interval. For simplicity of notation, we shall take it to be the unit interval [0, 1] without loss of generality.

Let C(0, 1) denote the class of continuous functions (with range in E) on the closed interval [0, 1]. Endowing it with the 'sup' norm, we know that it becomes a Banach space:

$$\|f\| = \sup |f(t)| , \quad 0 \leq t \leq 1$$

where $|\cdot|$ denotes the Euclidean norm. We note that it is a separable Banach space, and denote it by $\mathscr{C}$. By the Borel sets of $\mathscr{C}$ we mean the smallest sigma-algebra generated by all open sets. Let B be an m-dimensional Borel set in $E^{(m)}$. Then for arbitrary $t_1, t_2, \ldots t_n$, the sets in $\mathscr{C}$ defined by:

$$[f(\cdot) \in C \mid f(t_1), \ldots f(t_n) \in B]$$

are called 'cylinder sets' (and B is then referred to as the 'base'). We observe that the class of cylinder sets is a 'field'.

Lemma The class of Borel sets in $\mathscr{C}$ coincides with the smallest sigma-algebra generated by the class of cylinder sets.

Proof It is clear that cylinder sets are Borel sets. Conversely, since $\mathscr{C}$ is separable, any open set in $\mathscr{C}$ can be expressed as the union of a countable number of closed spheres; and every closed sphere, say with center $f_o$ and radius $d$ can be expressed:

$$\bigcap_n [f(\cdot) \mid |f(r_n) - f_o(r_n)| \leq d]$$

where $r_n$ denotes the countable collection of rational numbers in $[0,1]$. Hence open sets are contained in the smallest sigma-algebra generated by cylinder stes.

Note in particular that the Borel sets are generated as the smallest sigma-algebra containing sets of the form, $f(.) \in \mathscr{C}$:

$$[f(\cdot) \mid f(t) \in I, \; I \text{ an interval in } E].$$

We first note a trivial but useful fact which we state as a Lemma.

Lemma  Suppose $S(t;\omega)$, $0 \leq t \leq 1$, is a stochastic process such that sample functions are continuous with probability one. Then the mapping

$$\Psi(\omega) = S(\cdot;\omega)$$

mapping $\Omega$ into $\mathscr{C}$ is (Borel) measurable, and induces a probability measure on $\mathscr{B}_{\mathscr{C}}$ consistent with the finite dimensional distributions of the process.

Proof  Let $(\Omega, \mathscr{B}, p)$ denote the probability triple for the process $S(t;\cdot)$. Let us first prove that the mapping $\Psi(\cdot)$ is measurable. Let $U$ denote a Borel set in $E$, and let $C$ denote the cylinder set in $\mathscr{B}$ :

$$C = [f(\cdot) \in \mathscr{C} \mid f(t) \in U \; ; \; t \text{ fixed}]$$

Then

$$\Psi^{-1}(C) = [\omega | S(t;\omega) \in U]$$

and hence is measurable $\mathscr{B}$. Now consider the class of sets in such that inverse images in $\Omega$ are measurable. This is clearly a sigma-algebra and contains all cylinder sets. Hence it includes the sigma-algebra of Borel sets. Hence $\Psi(\cdot)$ is measurable. Defining the measure

$$p_S(A) = p(\Psi^{-1}(A))$$

we can readily see it is countably additive on $\mathscr{B}_{\mathscr{C}}$, and further agrees with the finite-dimensional distributions of the process.

Conversely, given any probability measure on $\mathscr{B}_{\mathscr{C}}$, we can define a corresponding stochastic process (with continuous sample paths) by:

$$S(t;\omega) = \omega(t)$$

where $\omega$ now denotes elements in $\mathscr{C}$.

It is rarely however that we are 'handed' a process with continuous sample paths. Almost always, this fact has to be deduced if possible from the finite dimensional distributions. Given a family of consistent finite-dimensional distributions, we can induce a finitely additive measure on the cylinder sets of $\mathscr{C}$ following (1.1). In general, however, this measure need not be extendable to be countably additive on the Borel sets of $\mathscr{C}$. But a sufficient condition for it is (2.1).

Theorem 2.0: Suppose (2.1) is satisfied. Then the finitely-additive measure induced on cylinder sets of $\mathscr{C}$ can be extended to be a countably additive probability measure on the sigma-algebra of Borel sets of $\mathscr{C}$.

Proof: We follow Parthasarathy [4].

First use Kolmogorov's theorem and go to $(X, \mathscr{S}, p(\cdot))$. Then for each n, define a mapping $\varphi_n(\cdot)$, mapping X into $\mathscr{C}$ by:

$$\varphi_n(x(\cdot)) = f(\cdot)$$

$$f(t) = x(m/2^n) + 2^n(t-m/2^n)\,(x(\overline{m+1}/2^n) - x(m/2^n))$$

$$\text{for } m/2^n \le t \le (m+1)/2^n, \quad 0 \le m < 2^n, \; m \text{ - integer.}$$

In other words we join the ordinates at the discrete timepoints $m/2^n$ by segments of straight lines. It is not difficult to see that $\varphi_n(\cdot)$ is measurable; that is to say, the inverse images of Borel sets:

$$\varphi_n^{-1}(B) = [x(\cdot) \in X;\ \varphi_n(x(\cdot)) \in B]$$

where B is a Borel set in $\mathscr{C}$, belong to $\mathscr{S}$. For this we have only to note that if B is of the form:

$$B = [f(\cdot) \in \mathscr{C},\ f(t_1) \in I\ ,\ \ I \text{ Borel set in } E]$$

then taking

$$m/2^n \le t_1 < (m+1)/2^n$$

we have that

$$\varphi_n^{-1}(B) = [x(\cdot) \in X,\ x(m/2^n) + 2^n(t_1-m/2^n)(x(\overline{m+1}/2^n) - x(m/2^n)) \in I]$$

which is clearly in $\mathscr{S}$. Hence the smallest sigma-algebra generated by inverse images of this form are in $\mathscr{S}$. But the smallest sigma-algebra generated by sets of the form B is indeed the class of all Borel sets. Hence $\varphi_n(\cdot)$ is measurable. Next we come to the crucial part; and show that $\varphi_n(x)$ is a Cauchy sequence in $\mathscr{C}$ for all x in X, except for a fixed set of measure zero. For this we begin with the bound:

$$\|\varphi_n(x) - \varphi_{n-1}(x)\| \leq \sup_{1 \leq k < 2^n} |x(k/2^n) - x(\overline{k-1}/2^n)| \tag{2.2}$$

This is evident from the fact that the maximum 'deviation' occurs at the subdivision points $(2k+1)/2^n$, and here the deviation is

$$|x(\overline{2k+1}/2^n) - x(2k/2^n) - (\tfrac{1}{2})(x(\overline{2k+2}/2^n) - x(2k/2^n))|$$

$$= |(\tfrac{1}{2})(x(\overline{2k+1}/2^n) - x(\overline{2k+2}/2^n)) + (\tfrac{1}{2})(x(\overline{2k+1}/2^n) - x(2k/2^n))|$$

Next we invoke the Chebychev inequality; and obtain:

$$p(\|\varphi_n(x) - \varphi_{n-1}(x)\| \geq \epsilon) \leq \frac{1}{\epsilon^r} \sum_{i=1}^{2^n} E(|x(i/2^n) - x(\overline{i-1}/2^n)|^r)$$

$$\leq \frac{k}{\epsilon^r} \sum_{i=1}^{2^n} (\frac{1}{2^n})^{1+\delta}$$

$$= \frac{k}{\epsilon^r} \frac{1}{2^{n\delta}}$$

Next let

$$0 < \theta < \delta/r$$

and let

$$A_n = [x \mid \|\varphi_n(x) - \varphi_{n-1}(x)\| \geq 2^{-n\theta}]$$

Then

$$p(A_n) \leq k\ 2^{nr(\theta - \delta/r)}$$

and since

$$\sum_{n=1}^{\infty} k\ 2^{nr(\theta - \delta/r)} < \infty$$

it follows (Borel-Cantelli Lemma):

$$p(\bigcap_{n=1}^{\infty} \bigcup_{j=n}^{\infty} A_j) = 0$$

Let

$$F = \bigcap_{n=1}^{\infty} \bigcup_{j=n}^{\infty} A_j$$

For any $x$ not in $F$ we have that, denoting the complement of $A_j$ by $\mathscr{C}\text{-}A_j$,

$$x \in \bigcup_{n=1}^{\infty} \bigcap_{j=n}^{\infty} (\mathscr{C} - A_j)$$

or, for each $x$ not in $F$,

$$\|\varphi_n(x) - \varphi_{n-1}(x)\| < 2^{-n\theta}$$

for <u>all</u> n sufficiently large (depending on $x$, of course). Hence for such $x$, for all $n$ sufficiently large and every $p$,

$$\|\varphi_n(x) - \varphi_{n+p}(x)\| \leq \sum_{n+1}^{n+p} 2^{-j\theta}$$

or,

$$\varphi_n(x) \text{ is a Cauchy sequence in } \mathscr{C}.$$

Denote the limit by $\varphi(x)$ for $x$ not in $F$ and define the limit to be zero (the zero function) on $F$. And define

$$\begin{aligned} \Psi(x) &= \varphi(x) \text{ on the complement of } F \\ &= 0 \text{ on } F \end{aligned}$$

Then $\Psi(\cdot)$ is measurable.
Note that for any $x(\cdot)$ not in $F$, we have that, if

$$\Psi(x(\cdot)) = f(\cdot),$$

then at each subdivision point $t_j$ (of the form $m/2^j$)

$$x(t_j) = f(t_j)$$

by construction. On the other hand, using (2.1) it follows that we have stochastic continuity; and hence if $t_k$ converges to $t$, a subsequence of $\{x(t_k)\}$ converges with probability 1 to $x(t)$. Hence for each $t$

$$x(t) = f(t)$$

with probability one where the exceptional set may depend on $t$. [In other words $(\Psi(x))(t)$ is equivalent to $x(t)$]. Next we define the probability measure on $\mathscr{C}$ by: for each Borel set $B$,

$$\hat{p}(B) = p(\Psi^{-1}(B))$$

Let us calculate the finite dimensional distributions corresponding to $\hat{p}$. Let I be an interval in E, and let

$$B = [f(\cdot) \varepsilon \mathscr{C}, \ f(t) \varepsilon I]$$

Then

$$\hat{p}(B) = p(\Psi^{-1}(B))$$

$$= p(x \varepsilon X \mid (\Psi(x))(t) \varepsilon I)$$

$$= p(x \mid x(t) \varepsilon I)$$

by the equivalence just proved. Hence $\hat{p}$ 'agrees' with the given finite dimensional distributions.

Remark Instead of beginning with a family of finite dimensional distributions, we could have begun equivalently with a stochastic process say $S(t;\omega)$, $t \varepsilon [0,1]$ and $(\Omega, \mathscr{B}, p)$ denoting the probability triple, such that the process satisfies (2.1):

$$E(\|S(t_2;\omega) - S(t_1;\omega)\|^r) \leq k|(t_2-t_1)|^{1+\delta}$$

The proof of Theorem 2.0 would then go over 'in toto' with $(\Omega, \mathscr{B}, p)$ in place of $(X, \mathscr{S}, p)$, with $S(t;\omega)$ in place of $x(t)$, and $\omega$ in place of $x$. We thus obtain a measurable mapping $\psi(\omega)$, mapping $\Omega$ into $\mathscr{C}$. Define the process $\tilde{S}(t;\omega)$ by

$$\psi(\omega) = \tilde{S}(\cdot\,;\omega)$$

$\tilde{S}(t;\omega)$ denoting the value of the function $\psi(\omega)$ at t. Then the process $\tilde{S}(t;\omega)$ has continuous sample function with probability one, and further

$$\tilde{S}(t;\omega) = S(t;\omega)$$

with probability one, for each t. The process $\tilde{S}(t;\omega)$ is thus equivalent to $S(t;\omega)$.

Example

Perhaps the simplest example of a process satisfying (2.1) can be constructed in the following way. Let $x(t;\omega)$ be a separable stochastic process with finite second moments, with zero mean, and such that the covariance function:

$$E(\, x(t;\omega)\, x(s:\omega)^{*}\,) = r(t;s)$$

is continuous on $0 \leq s, t \leq 1$. The process is then stochastically continuous, and as we have noted $x(t;\omega)$ is jointly measurable in $\omega$ and $t$ (Lebesgue measure in t). Now, by Schwarz inequality, we have for each t:

$$\int_{\Omega} \int_0^t \|x(s;\omega)\| \, ds \, dp \leq \sqrt{t} \sqrt{\int_{\Omega} \int_0^t \|x(s;\omega)\|^2 \, ds \, dp} = \sqrt{\int_0^t \mathrm{Tr.}\, r(s;s)\, ds} \cdot \sqrt{t} < \infty$$

Hence by Fubini's Theorem

$$\int_0^t x(s;\omega) ds$$

is defined for almost all $\omega$ as a Lebesgue integral, and further is measurable in $\omega$. Let us define:

$$S(t;\omega) = \int_0^t x(s;\omega)\, ds$$

for each $t$, $0 \leq t \leq 1$. Note that the exceptional set where $S(t;\omega)$ is not defined may depend on t. Nevertheless $S(t;\omega)$ defines a stochastic process. Next for any two points $t_1, t_2$, we have, omitting a $\omega$-set of measure zero:

$$S(t_2;\omega) - S(t_1;\omega) = \int_{t_1}^{t_2} x(s;\omega)\, ds$$

and hence it follows that

$$E(\, \|(S(t_2;\omega) - S(t_1;\omega)\|^2\,) = \int_{t_1}^{t_2} \int_{t_1}^{t_2} \mathrm{Tr.}\, r(s;\sigma)\, ds\, d\sigma$$

$$\leq \max |\mathrm{Tr.}\, r(s;t)| \cdot (t_2 - t_1)^2$$

so that (2.1) is satisfied with $r = 2$, $\delta = 1$. Hence Theorem 2.0 applies, and there is an equivalent process $\tilde{S}(t;\omega)$ such that it has continous sample paths with probability one, and

$$\tilde{S}(.\,;\omega) = \psi(\omega)$$

defines a measurable function mapping $\Omega$ into $\mathscr{C}$ . Finally, we note that given any continuous covariance function $r(t;s)$ we can always construct a Gaussian process $x(t;\omega)$. In that case, $S(t;\omega)$ is of course also Gaussian.

Let us also note, before we leave the subject, that not every process with continuous sample paths need satisfy condition (2.1). Here is a trivial example. Let $g(\cdot)$ be an example in $\mathscr{C}$. Define a measure on the Borel sets of $\mathscr{C}$ by:

$$p(B) = 1 \text{ if } B \text{ contains the element } g$$

$$= 0 \text{ otherwise}$$

This is clearly a countably additive probability measure. Now take

$$g(t) = (1/\text{Log } t)\, v \qquad 0 \leq t \leq 1/2$$

$$= (1/\text{Log}(1/2))v \quad 1/2 \leq t \leq 1$$

where $v$ is a unit vector. Taking $\omega$ to be any element of $\mathscr{C}$, define

$$S(t;\omega) = \omega(t)$$

Then the process $S(t;\omega)$ obviously continuous sample paths.

However (2.1) is not satisfied since

$$E(\| S(t;\omega) - S(0;\omega)\|^r) = |1/\text{Log } t|^r$$

and

$$\sup_{0 \leq t \leq 1/2} \frac{1}{(t^{1+\delta})(|\text{Log } t|^r)}$$

is unbounded for all $r > 0$ and all $\delta > 0$. In this volume, as we have noted, we only deal with processes for which (2.1) is satisfied. For a non-trivial example, see p. 28.

Wiener Process

An important special case of measures on $\mathcal{C}$, for us the one central case, is Wiener measure.

Theorem 2.1 There exists a measure W on the Borel sets of $\mathcal{C}$ (which we shall denote henceforth by $\mathcal{B}_{\mathcal{C}}$) such that

$$P_W[f(\cdot) \mid f(0) = 0] = 1$$

and for any finite number of indices $t_1 < t_2 < \ldots < t_m$,

$$P_W[f(\cdot) \mid f(t_k) \in I_k, \quad k = 1, \ldots m]$$

where $I_k$ are intervals in E, is given by

$$\int_{I_1} \cdots \int_{I_m} G(x_1, \ldots x_m) \, d|x_1| \ldots d|x_m|$$

where $G(\ldots.)$ is mn-variate Gaussian with zero mean and

$$\int \cdots \int x_i x_j^* \, G(x_1, \ldots, x_m) \, d|x_1| \ldots d|x_m| = \min(t_i, t_j) I_n$$

Proof It is readily verified that the given joint distributions satisfy (2.1) with $r = 4$, $k = n(n+2)$, and $\delta = 1$, and satisfy the consistency requirements.

Remark We recall now that any measure can be 'completed' [2] that is to say, we consider the class of subsets of measurable sets of measure zero and define the measure to be zero on them, and the 'complete' measure is then defined on the (larger) sigma-algebra of sets which differ from the original sigma-algebra by sets of measure zero. It is in this sense that we shall talk about 'sets of Wiener measure zero'.

With $\omega$ denoting 'points' in $\mathscr{C}$ and defining

$$W(t;\omega) \quad 0 \leq t \leq 1$$

to be the 'function-value' at t corresponding to $\omega$, the measure $P_W$ being the Wiener measure above, we have a 'Wiener process' which is thus a Gaussian function-space process, with sample space $\mathscr{C}$. We note that for $t_1 < t_2 < t_3$,

$$W(t_2;\omega) - W(t_1;\omega)$$

is Gaussian with mean zero and covariance $(t_2-t_1) I_{(n)}$, $I_{(n)}$ being the $n x n$ identity matrix and is independent of

$$W(t_3;\omega) - W(t_2;\omega).$$

Hence $W(t;\omega)$ is also a Gaussian process with independent increments such that

$$E((W(t_2;\omega) - W(t_1;\omega))(W(t_2;\omega) - W(t_1;\omega))^*) = (t_2-t_1) I_{(n)}$$

and has sample functions which are continuous. It is also readily verified that

$$R(t;s) = E(W(t;\omega) W(s;\omega)^*) = (\min (s,t) I_{(n)} \tag{2.3}$$

Measurability For each t, $W(t;\omega)$ is actually continuous in $\omega$, being a continuous linear functional on $\mathscr{C}$. Being thus continuous in both variables, t and $\omega$, it is measurable jointly with respect to $\omega$ and t (Lebesgue measurable sets in $[0,1]$).

Let $\phi(t)$ be any function in $\mathscr{C}$. For each $\omega$, $W(t;\omega)$ is a continuous function of t. Hence we can define the integral

$$\int_0^1 \phi(t)^* W(t;\omega)dt$$

as a Riemann-integral for every $\omega$. Moreover, it defines actually a continuous linear functional on $\mathscr{C}$. In fact

$$\int_0^1 | \phi(t)^* (W(t;\omega_1) - W(t;\omega_2)) | dt$$

$$\leq \left( \sup_{0 \leq t \leq 1} | \phi(t) | \right) \| \omega_1 - \omega_2 \|$$

since

$$|W(t;\omega_1) - W(t;\omega_2)| \leq \|\omega_1 - \omega_2\|$$

The integral defines a Gaussian of mean zero and with variance

$$\int_0^1 \int_0^1 \phi(t)^* R(t;s) \phi(s) \, ds \, dt$$

where $R(t,s)$ is defined by (2.3).

Wiener Measure: Extension to Infinite Interval

For most of our work it is enough to consider the Wiener process defined on a compact interval. However, for some purposes, as for example, in asymptotic theory where we need to allow time to become infinite in limiting operations, we need the Wiener process defined on $0 \leq t < \infty$. The appropriate class of sample functions is the class of

+May be omitted on first reading; uses " locally convex spaces ".

functions $f(t)$, $0 \leq t < \infty$, continuous on compact subintervals. This is of course a linear space. We topologize it by the family of seminorms $\|\cdot\|_T$, where

$$\|f\|_T = \underset{0 \leq t \leq T}{\text{Max}} \|f(t)\| \quad , \quad 0 < T < \infty$$

Thus topologized we obtain a locally convex space, separable and complete, but not a Banach space any more. We denote it by $C[0, \infty]$. Let us note that $f_n(\cdot)$ is a Cauchy sequence in this space if and only if it is a Cauchy sequence for every seminorm, or equivalently for a sequence of seminorms $\|\cdot\|_{T_i}$ where

$$T_i < T_{i+1} \quad ; \quad T_i \longrightarrow \infty$$

We define cylinder sets the same way as before. The Borel sets are again the sigma-algebra of sets generated by the class of all open sets. We also have:

Lemma: The class of Borel sets in $C[0, \infty)$ coincides with the sigma algebra generated by the cylinder sets.

Proof: Obviously every cylinder set is a Borel set. Conversely, let $0$ be an open set in $C[0, \infty)$. The space being separable, actually 'countably normed', we have that $0$ can be expressed as a countable union of closed spheres of the form:

$$[f| \ \|f(\cdot) - f_i(\cdot)\|_{T_i} \leq r_i]$$

for a countable index i. For fixed i, any closed sphere is as before in the sigma algebra generated by cylinder sets.

Next let us show that under condition (2.1) we can induce a corresponding countably additive measure on the Borel sets. As before (as in the case of $C[0,1]$), let $X$ denote the class of all functions on $[0,\infty)$, $S$ the sigma-algebra generated by cylinder sets and $p(\cdot)$ the measure induced by the given finite dimensional distributions via the Kolmogorov extension theorem. Then for each $n$ define the mapping $\varphi_n(\cdot)$ on $X$ into $C[0,\infty)$, by:

$$(x(\cdot)) = f(\cdot);$$

$$f(t) = x(m/2^n) + 2^n(t-m/2^n)(x(\overline{m+1}/2^n) - x(m/2^n))$$

$$\text{for} \quad m/2^n \leq t \leq \overline{m+1}/2^n \;; \quad m \text{ integer} \geq 0.$$

By virtually the same argument as before, $\varphi_n(x)$ is measurable. We next need to show that $\varphi_n(x)$ defines a Cauchy sequence in $C[0,\infty)$, for each $x(\cdot)$, omitting a set of measure zero. But for each seminorm $\|\cdot\|_T$, it is as though we are considering $C[0,T]$, and hence the same argument as before goes through. The remainder of the argument goes through as before also.

We can thus in particular define a Wiener measure $p_W(\cdot)$ say on $C[0,\infty)$ such that

$$p_W([f| \; f(0) = 0]) = 1$$

$$W(t;\omega) = \omega(t)$$

$$E(W(t;\omega)W(s;\omega)^*) = (\min s, t) \; I_m$$

where $m$ is the dimension of the process.

Finally, we note that in a similar manner we can induce a Wiener measure on $C(-\infty, +\infty)$, the locally convex space of functions continuous on compact subintervals with seminorms $\|\cdot\|_{[a,b]}$ defined by:

$$\|f\|_{[a,b]} = \text{Max } \|f(t)\|, \quad a \leq t \leq b.$$

$$E(W(t;\omega)\, W(s;\omega)^*) = (\min |t|, |s|)\, I_m$$

and again:

$$P_W([f \mid f(0) = 0]) = 1$$

## 2.2 Stochastic Integrals: Linear Case

Let $\phi(t)$ be any function in $L_2(0,1)$ over $E$. We wish now to define the stochastic integral (with respect to the Wiener process)

$$\int_0^1 [\phi(t), dW(t;\omega)]$$

First let $\phi(t)$ be a step-function of the form:

$$\phi(t) = v_i \quad t_i \leq t < t_{i+1}, \quad i = 0, \ldots n-1;$$

$$t_0 = 0 \leq t_1 \leq t_2 \ldots \leq t_n = 1$$

corresponding to an arbitrary subdivision of $[0,1]$ into subintervals. Then we define

$$\int_0^1 [\phi(t), dW(t;\omega)] = \sum_{i=0}^{n-1} \left[ v_i (W(t_{i+1};\omega) - W(t_i;\omega)) \right] \tag{2.3a}$$

This defines a Gaussian random variable with mean zero and variance:

$$\sum_{i=0}^{n-1} \|v_i\|^2 (t_{i+1} - t_i) = \int_0^1 \|\phi(t)\|^2 \, dt = \|\phi(.)\|^2$$

[The integral of course defines a continuous linear functional on $C$, and the norm of the functional is

$$\leq 2 \sum_{0}^{n-1} \|v_i\| ]$$

Let $L_2(\mathscr{C})$ denote the $L_2$ space of all random variables (measurable $\mathscr{B}_{\mathscr{C}}$) with finite second moment with respect to Wiener measure. Then (2.3a) defines a mapping into $L_2(\mathscr{C})$. It is readily verified to be linear on the linear subspace of step functions of the type considered. But this subspace is dense in $L_2(0, 1)$. If we denote the mapping by $\mathscr{L}$, we have

$$E\,|(\mathscr{L}(\phi(\cdot)))|^2 = \|\phi(\cdot)\|^2 \quad \ldots . \tag{2.4}$$

and hence can be extended to be continuous on $L_2(0, 1)$. In other words if $\phi(\cdot)$ is any element in $L_2(0, 1)$ we know that we can find a sequence $\phi_n(\cdot)$ of step functions of the type considered which converge to $\phi(\cdot)$, and since $\{\phi_n(\cdot)\}$ is then Cauchy sequence in $L_2(0, 1)$, so is

$$\mathscr{L}(\phi_n(\cdot))$$

in $L_2(\mathscr{C})$. Since $L_2(\mathscr{C})$ is complete it follows that we can define

$$\mathscr{L}(\phi(\cdot)) = \underset{n}{\text{limit}} \ (\text{in } L_2(\mathscr{C})) \ \mathscr{L}(\phi_n(\cdot))$$

The limit is of course independent of the particular sequence chosen, thanks again to (2.4), and moreover a subsequence of $\mathscr{L}(\phi_n(\cdot))$ will converge to the limit a.e., and in particular for this subsequence:

$$E[e^{it\mathscr{L}(\phi_n(\cdot))}] \to E[e^{it\mathscr{L}(\phi(\cdot))}] \tag{2.5}$$

$$E\,|\mathscr{L}(\phi(\cdot))|^2 = \|\phi(\cdot)\|^2$$

and further, if $\phi(\cdot)$, $\Psi(\cdot)$ are any two elements in $L_2(0,1)$ we have

$$E([\mathscr{L}(\phi(\cdot)), \mathscr{L}(\Psi(\cdot))]) = [\phi(\cdot), \Psi(\cdot)]$$

where the right side indicates inner product in $L_2(0,1)$. It is hardly necessary to add that $\mathscr{L}$ is a linear continuous (isometric) transformation of $L_2[0,1]$ into $L_2(\mathscr{C})$ and (from (2.5))

$$\mathscr{L}(\phi(\cdot))$$

is Gaussian with zero mean and variance

$$= \| \phi(\cdot) \|^2$$

If $\phi(\cdot)$ is actually continuous then we can approximate $\mathscr{L}(\phi(\cdot))$ by sums of the form, with $\{t_i\}$ as before,

$$\sum_{i=0}^{n-1} [\phi(\tau_i), (W(t_{i+1};\omega) - W(t_i;\omega))]$$

where $t_i \le \tau_i \le t_{i+1}$, since the difference (between this sum and the integral)

$$= \sum_{i=0}^{n-1} \int_{t_i}^{t_{i+1}} [\phi(\tau_i) - \phi(t), dW(t;\omega)]$$

and has variance:

$$\sum_0^{n-1} \int_{t_i}^{t_{i+1}} |\phi(\tau_i) - \phi(t)|^2 dt \to 0 \quad \text{as} \quad \max_i |t_{i+1} - t_i| \to 0 .$$

If $\phi(\cdot)$ is absolutely continuous with derivative in $L_2[0,1]$, we can 'integrate by parts':

$$\int_0^1 [\phi(t), dW(t;\omega)] = [\phi(1), W(1)] - \int_0^1 [\phi'(t), W(t;\omega)]dt \quad \text{with pr. one}$$

This follows from the calculation:

$$\sum_{i=0}^{n-1} [\phi(t_i), W(t_{i+1};\omega) - W(t_i;\omega)]$$

$$= - [\phi(0), W(0;\omega)] - \sum_1^n [\phi(t_i) - \phi(t_{i-1}), W(t_i;\omega)]$$

$$+ [\phi(1), W(1;\omega)]$$

and

$$\sum_1^n [\phi(t_i) - \phi(t_{i-1}), W(t_i;\omega)] = \sum_1^n \int_{t_{i-1}}^{t_i} [\phi'(\sigma), W(t_i;\omega)]d\sigma$$

$$\longrightarrow \int_0^1 [\phi'(t), W(t;\omega)]dt$$

Series Expansions:

Let $\phi_n(\cdot)$ be a complete orthonormal sequence in $L_2(0,1)$. Let

$$\zeta_n(\omega) = [\phi_n(t), dW(t;\omega)]$$

Then the random variables $\zeta_n(\omega)$ are zero mean, unit variance Gaussians which are mutually independent. For any $\phi(\cdot)$ in $L_2(0,1)$ we have the series expansion:

$$\int_0^1 [\phi(t), dW(t;\omega)] = \sum_1^\infty \zeta_n(\omega) [\phi(\cdot), \phi_n(\cdot)]$$

where the series converges in the mean of order two, and with probability one. The convergence in the mean square follows from the fact that:

$$E\left(\left|\int_0^1 [\phi(t), dW(t;\omega)] - \sum_1^m \zeta_n(\omega)[\phi(\cdot), \phi_n(\cdot)]\right|^2\right)$$

$$= [\phi(\cdot), \phi(\cdot)] - \sum_1^m [\phi(\cdot), \phi_n(\cdot)]^2$$

$$\to 0$$

A ready application of the Kolmogorov inequality (Corollary, Theorem 1.1) yields convergence with probability one. As a special case, we can find a useful series expansion for the Wiener process itself (see Shepp [5]):

$$W(t;\omega) = \sum_1^\infty \zeta_n(\omega) \int_0^t \phi_n(s)ds, \quad 0 \leq t \leq 1 \tag{2.6}$$

The series converges again in the mean square as well as with probability one. To see this, let us fix $t$, and let $v$ be any vector in $E$, and define the function $\phi(s)$, by:

$$\begin{aligned} \phi(s) &= v \qquad 0 \leq s \leq t \\ &= 0 \qquad t < s < 1 \end{aligned}$$

Clearly $\phi(\cdot)$ is in $L_2(0,1)$, and hence we have:

$$[\phi(\cdot), \phi_n(\cdot)] = [v, \int_0^t \phi_n(s)ds]$$

so that

$$\sum_1^\infty [v, \int_0^t \phi_n(s)ds]^2 = [\phi(\cdot), \phi(\cdot)] = t[v, v]$$

Hence, if $e_k$, $k = 1,..n$, is an orthonormal basis in E, we have

$$\sum_{j=1}^{\infty} \| \int_0^t \phi_j(s)ds \|^2 = \sum_{j=1}^{\infty} \sum_{1}^{n} [e_k, \int_0^t \phi_j(s)ds]^2$$

$$= \sum_{k=1}^{n} \sum_{j=1}^{\infty} [e_k, \int_0^t \phi_j(s)ds]^2$$

$$= \sum_{k=1}^{n} t[e_k, e_k] = n\,t$$

Hence the right side of (2.6) converges in the mean square; and with probability one by Kolmogorov inequality as before. Finally,

$$[v, W(t;\omega) - \sum_{1}^{\infty} \zeta_j(\omega) \int_0^t \phi_j(s)ds]$$

$$= \int_0^1 [\phi(s), dW(s;\omega)] - \sum_{1}^{\infty} \zeta_j(\omega) [\phi(\cdot), \phi_j(\cdot)]$$

$$= 0,$$

and v being arbitrary, (2.6) follows.

For our purposes the expansion (2.6) is more useful than the Karhunen-Loeve expansion (1.3) obtained earlier.

**Problem**

$$\int_0^1 E(\| W(t;\omega) - \sum_{1}^{m} \zeta_k(\omega) \int_0^t \phi_k(s)\, ds \|^2)\, dt \longrightarrow 0$$

**Hint:**

$$E \| W(t;\omega) - \sum_{1}^{m} \zeta_k(\omega) \int_0^t \phi_k(s)ds \|^2 = (nt - \sum_{1}^{m} \| \int_0^t \phi_k(s)\, ds \|^2) \geq 0;$$

$nt = E[\| W(t;\omega) \|$

**and so,**

$$\sum_{1}^{m} \| \int_0^t \phi_k(s)ds \|^2 \text{ converges boundedly to } (nt) \text{, in } L_1[0,1].$$

Linear Stochastic Equations

Often, in the applications, we need to interpret differential equations with "white noise" forcing term:

$$\frac{d}{dx} x(t;\omega) = A(t)\, x(t;\omega) + B(t)\, n(t;\omega)$$

where $n(t;\omega)$ is the "derivative of the Wiener process". However, strictly speaking, the sample functions of a Wiener process are not differentiable. In fact they are not of bounded variation with probability one (see Doob [2 ]). To make sense of the above equation, we rewrite it in the 'integral' form:

$$x(t;\omega) = \xi(\omega) + \int_0^t A(s)\, x(s;\omega)ds + \int_0^t B(s)\, dW(s;\omega)$$

We have thus a stochastic linear (integral) equation. We wish now to consider the problem of existence and uniqueness of solutions to such equations since it plays an essential role in Filtering and Control theories. First, however, we have to discuss the simplest such equation:

$$x(t;\omega) = \int_0^t B(s)dW(s;\omega)$$

Let $F(s)$ be an $m \times n$ matrix function, Lebesgue measurable on $[0,1]$ and such that

$$\int_0^1 \|F(s)\|^2\, ds < \infty$$

Let

$$S(t;\omega) = \int_0^t F(s)\, dW(s;\omega) \qquad 0 \le t \le 1 \tag{2.7}$$

which we know defined for each $t$, for almost every $\omega$. However it is not defined for every $\omega$, and in particular, the exceptional set of points $\omega$ on which it is not defined may well depend on $t$. We wish now to rectify this situation.

Theorem 2.2 The integral in (2.7) can be determined so that it is continuous in $t$, $0 \le t \le 1$ for almost all $\omega$.

Proof What we shall prove is that we can determine an equivalent process $\tilde{S}(t;\omega)$ with continuous sample paths. If $F(s)$ is absolutely continuous with square-integrable (on $[0,1]$) derivative we can do this very simply. For then,

$$S(t;\omega) = \int_0^t F(s)dW(s;\omega) = [F(t), W(t;\omega)] - \int_0^t F'(s)W(s;\omega)ds$$

with probability one, and since the right side is defined for every $\omega$, we may just define that to be $\tilde{S}(t;\omega)$ for every $\omega$. $\tilde{S}(t;\omega)$ is then continuous in $t$, $0 \leq t \leq 1$, for every $\omega$.

If $F(s)$ is bounded in $[0,1]$, then we can use (2.1):(see Problem below). How[illegible] condition (2.1) need not be satisfied in general. For example, take

$$F(s) = v \left( \sqrt{d/ds\ (-1/\operatorname{Log}(s/2))} \right)$$

where $v$ is a unit vector, so that $S(t;\omega)$ is Gaussian with variance:

$$(-1)(1/\operatorname{Log}(t/2)).$$

To handle the general case, we use the following bound:

Lemma Suppose $S(t;\omega)$ in (2.7) is determined as a continuous function of $t$ for almost every $\omega$. Then:

$$P_W \left[ \sup_{0 \leq t \leq 1} \left\| \int_0^t F(s)dW(s;\omega) \right\| > \epsilon \right] \leq \frac{1}{\epsilon^2} \int_0^1 \|F(s)\|^2 ds \tag{2.8}$$

Proof

Let

$$A = \left[ \omega \,\middle|\, \sup_{0 \leq t \leq 1} \left\| \int_0^t F(s)\, dW(s;\omega) \right\| > \epsilon \right]$$

Then by the assumed continuity of $S(t;\omega)$ in $t$, we observe that $A$ is measurable. In fact if, for each integer $k$, we define:

$$A_n(k) = \left[ \omega \,\middle|\, \sup_{0 \leq m \leq 2^n} \left\| \int_0^{m/2^n} F(s)\, dW(s;\omega) \right\| \geq \epsilon - \frac{1}{k} \right], \text{ m-integer}$$

then $A_n(k)$ is clearly monotone (non-decreasing) in $n$ for each $k$, and

$$A = \bigcap_k \bigcup_n A_n(k)$$

But the integral over non-overlapping intervals being independent, a simple application of the Kolmogorov inequality shows that

$$P_W(A_n(k)) \leq \frac{1}{(\epsilon-\frac{1}{k})^2} \int_0^1 \|F(s)\|^2 \, ds$$

from which the Lemma follows.

Next let us recall that if $F(\cdot)$ is any element in $L_2(0, 1)$, we can find a sequence $F_n(\cdot)$ of functions which are absolutely continuous with derivative in $L_2(0, 1)$ such that

$$\int_0^1 \|F(s) - F_n(s)\|^2 \, ds \longrightarrow 0$$

Using the notation:

$$\|F_n(\cdot) - F_m(\cdot)\|^2 = \int_0^1 \|F_n(s) - F_m(s)\|^2 ds$$

$$\|f(\cdot;\omega)\| = \sup_{0 \leq t \leq 1} \|f(t;\omega)\|$$

$$S_n(t;\omega) = \int_0^t F_n(s) \, dW(s;\omega)$$

the Lemma yields:

$$P_W[\|S_n(\cdot;\omega) - S_m(\cdot;\omega)\| \geq \epsilon] \leq \|F_n(\cdot) - F_m(\cdot)\|^2/\epsilon^2$$

We have thus a Cauchy ('fundamental') sequence in probability. Hence we can find a subsequence that converges with probability one. Although this is proved in standard texts (e.g., Loeve [10], Gikhman-Skorokhod [3]) for the case of random variables with range in a Euclidean space, essentially the same proof goes through in the present case even though the variables have their range in $\mathscr{C}$. To remove any doubt, we append a short proof.

Let $\theta$ be fixed, $0 < \theta < 1$, and choose integers $n_k$ such that for each integer $k$:

$$\| F_n(\cdot) - F_m(\cdot) \|^2 \leq 2^{-k\theta} \quad \text{for all } n, m \geq n_k.$$

Clearly, we can choose the sequence $n_k$ to be non-decreasing. Next choose any $r$ such that $0 < r < \theta < 1$ and let

$$B_k = [\omega \,|\, \| S_{n_k}(\cdot;\omega) - S_{n_{k+1}}(\cdot;\omega) \| > 2^{-kr/2}]$$

$$S_k'(\cdot;\omega) = S_{n_k}(\cdot;\omega)$$

and

$$\Lambda = \bigcap_{m=1}^{\infty} \bigcup_{k=m}^{\infty} B_k$$

Then since

$$\mathcal{P}_W(B_k) \leq 2^{-k(\theta-\gamma)}$$

we have (Borel Cantelli lemma):

$$\mathcal{P}_W(\Lambda) = 0$$

Moreover for any $\omega$ not in $\Lambda$, there exists an integer $k(\omega)$ such that for all $k \geq k(\omega)$:

$$\| S_{n_k}(\cdot;\omega) - S_{n_{k+1}}(\cdot;\omega) \| \leq 2^{-k\gamma/2}$$

Next let $\epsilon > 0$ be given. Choose $k(\epsilon)$ such that for all $k \geq k(\epsilon)$:

$$(2)(2^{-k\gamma/2})/(1-2^{-\gamma/2}) < \epsilon$$

Let us take any $\omega$ not in $\Lambda$, and let:

$$n(\epsilon;\omega) = \max\left\{k(\epsilon),\ k(\omega)\right\},$$

Then for all $n, m > n(\epsilon;\omega)$, we have that:

$$\| S'_n(\cdot;\omega) - S'_m(\cdot;\omega) \| < \epsilon$$

For, we have only to note that

$$\| S'_n(\cdot;\omega) - S'_m(\cdot;\omega) \| \leq \| S'_n(\cdot;\omega) - S_{n_k}(\cdot;\omega) \| + \| S_{n_k}(\cdot;\omega) - S'_m(\cdot;\omega) \|$$

$$\leq 2^{-k\gamma/2}/(1-2^{-\gamma/2}) + \text{same}$$

$$< \epsilon \text{ by construction}$$

Hence $S'_n(\cdot;\omega)$ converges (in $\mathscr{C}$-norm) for each $\omega$ not in $\Lambda$. Let us denote the limit, by $\tilde{S}(\ ;\omega)$. Then $\tilde{S}(t;\omega)$ is continuous in $t$ for almost every $\omega$. Moreover

$$E(\| \tilde{S}(t;\omega) - S(t;\omega) \|^2) = 0$$

so that

$$S(t;\omega) = \tilde{S}(t;\omega)$$

omitting at most a fixed set of measure zero, and thus we have produced a continuous equivalent version of $S(t;\omega)$.

Problem

Suppose $F(\cdot)$ is essentially bounded on $(0,1)$. Then show that (2.1) is satisfied and hence use Parathasarathy's construction to obtain a continuous version of $S(t;\omega)$.

HINT:

$$\begin{aligned} E[|S(t_1;\omega) - S(t_2;\omega)|^4] &= 3\,[E|\int_{t_1}^{t_2} F(s)d\,W(s;\omega)|^2]^2 \\ &= 3\left[\int_{t_1}^{t_2} \|F(s)\|^2\, ds\right]^2 \\ &\leq 3\, m^4(t_2-t_1)^2 \end{aligned} \qquad (2.$$

where

$$m = \text{ess.} \sup_{0\leq s\leq 1} \|F(s)\|.$$

Let us now consider the problem of solving the following linear stochastic integral equation:

$$x(t;\omega) = \xi(\omega) + \int_0^t A(s)\, x(s;\omega)ds + \int_0^t B(s)dW(s;\omega), \quad 0\leq t\leq L<\infty \qquad (2.10$$

where $\xi(\omega)$ is a given random variable, independent of $W(\cdot;\omega)$.

$A(s)$ is a Lebesgue measurable m-by-m matrix function

$B(s)$ is a Lebesgue measurable m-by-n matrix function

and

$$\int_0^L \|A(s)\|^2 \, ds + \int_0^L \|B(s)\|^2 \, ds < \infty \qquad (2.11)$$

Sometimes (2.10) is written in the differential form:

$$dx(t;\omega) = A(s)\, x(t;\omega)\, dt + B(t)\, dW(t;\omega);\ x(0;\omega) = \xi(\omega)$$

but this is to be looked upon only as a shorthand notation for (2.10). What shall we mean by a solution of (2.10)? Minimally, any stochastic process on $[\mathscr{C}, \mathscr{B}_{\mathscr{C}}, W]$, for which the integrals in (2.10) can be defined. We can actually do a little better. By a 'continuous solution' of (2.10) we shall mean a stochastic process such that $x(t;\omega)$ is continuous in $0 \le t \le L$, and satisfies (2.10) for every $t$ in $[0, L]$, omitting a fixed ω-set of measure zero. We begin with uniqueness.

Lemma Suppose $x_1(t;\omega)$, $x_2(t;\omega)$ are two continuous solutions of (2.10). Then

$$x_1(t;\omega) = x_2(t;\omega)$$

for every $t$ except for a fixed $\omega$-set of measure zero. In fact

$$P_W \left[ \sup_{0 \le t \le L} \|x_1(t;\omega) - x_2(t;\omega)\| = 0 \right] = 1$$

Proof

Let

$$y(t;\omega) = x_1(t;\omega) - x_2(t;\omega)$$

Then $y(t;\omega)$ is a continuous solution of:

$$y(t;\omega) = \int_0^t A(s)\, y(s;\omega)\, ds \quad \ldots\ldots. \tag{2.12}$$

for $\omega$ not in $\Lambda$ say, where $\Lambda$ is a set of measure zero. But for each $\omega$ not in $\Lambda$, (2.12) is a 'deterministic' equation, and readily yields that

$$y(t;\omega) = 0 \quad \text{for every } t \text{ and } \omega \bar{\epsilon} \Lambda.$$

<u>Lemma</u> (2.10) has a continuous solution.

<u>Proof</u> Let $\Phi(t)$ be the fundamental matrix solution of

$$\dot{\Phi}(t) = A(t)\; \Phi(t) \;\; ; \;\; \Phi(0) = I$$

Then let

$$S(t;\omega)$$

be a continuous version of

$$\int_0^t \Phi(s)^{-1}\, B(s)\, dW(s;\omega), \quad 0 \le t \le L$$

and finally let

$$x(t;\omega) = \Phi(t)(S(t;\omega) + \xi(\omega))$$

which is then a continuous function of $t$ excepting for a set $\Lambda$ of measure zero. Let us prove that this is a solution of (2.10).

For this note that

$$\int_0^t B(s)\, dW(s;\omega)$$

$$= \int_0^t \Phi(s)\; \Phi(s)^{-1}\, B(s)\, dW(s;\omega)$$

and integrating by parts, we get:

$$= \Big[ \phi(s)\, S(s;\omega) \Big]_0^t - \int_0^t \left(\frac{d}{ds}\,\phi(s)\right) S(s;\omega)\, ds$$

This equality can clearly be defined to hold for every $\omega$ not in $\Lambda$. But for $\omega$ not in $\Lambda$,

$$\int_0^t A(s)\, x(s;\omega)\, ds = \int_0^t \left(\frac{d}{ds}\,\phi(s)\right)\Big( S(s;\omega) + \xi(\omega) \Big)\, ds$$

$$= \phi(t)(S(t;\omega) + \xi(\omega)) - \int_0^t B(s)dW(s;\omega) - \xi(\omega)$$

$$= x(t;\omega) - \int_0^t B(s)\, dW(s;\omega) - \xi(\omega)$$

as required.

We have thus established that (2.10) has a unique (within equivalence), continuous solution.

Problem

Let $x(t;\omega)$ be a (continuous) solution of (2.10) and let

$R(t) = E(x(t;\omega)x(t;\omega)^*]$

show that $R(t)$ is absolutely continuous with

$$R(t) = A(t)\, R(t) + R(t)\, A(t)^* + B(t)\, B(t)^* \text{ a.e.} \qquad (2.13)$$

Hint

$$R(t) = \Phi(t)\, E[\xi(\omega)\, \xi(\omega)^*]\, \Phi(t)^*$$

$$+ \Phi(t) \int_0^t \Phi(s)^{-1}\, B(s)\, B(s)^*\, \Phi(s)^{-1*} ds\, \Phi(t)^*$$

Stochastic Differential Systems

We are now in a position to define the class of stochastic processes that we shall deal with for the most part in this book. Because of the structure imposed, we also refer to it as a (linear) 'stochastic differential system'. Thus let:

$$Y(t;\omega) = \int_0^t C(s)x(s;\omega)ds + \int_0^t D(s)dW(s;\omega), \quad 0 \leq t \leq L \tag{2.14}$$

where $x(t;\omega)$ is defined by the stochastic equation (2.10),

$C(s)$ is q-by-m and continuous on $[0, L]$

$D(s)$ is q-by-n and continuous on $[0, L]$

and for simplicity we shall take $\xi(\omega) = 0$, so that $x(0;\omega) = 0$. Then (2.10), (2.14) together define a 'stochastic differential system'.

Note that $Y(t;\omega)$ can be defined so that it is continuous in $t$, $0 \leq t \leq L$, excepting an $\omega$-set of (Wiener) measure zero. Furthermore, we can define a stochastic integral with respect to the $Y(t;\omega)$ process. Let $f(\cdot)$ be a function in $L_2(0, L)^q$ ; we first define the integral for a-step function taking on a finite number of values each on a subinterval, as before. We note that for a step function the following equality holds:

$$\int_0^L [f(t), dY(t;\omega)] = \int_0^L [f(t), C(t)x(t;\omega)]\, dt + \int_0^L [f(t), D(t)\, dW(t;\omega)] \tag{2.15}$$

Now if we take a Cauchy sequence of step functions, the right-side converges, and hence so does the left-side. And the stochastic integral is then defined as this limit. Hence finally (2.15) holds for every $f(\cdot)$ in $L_2[0, L]^q$. This may be indicated in differential notation as:

$$dY(t;\omega) = C(t)\, x(t;\omega)\, dt + D(t)\, dW(t;\omega) \tag{2.16}$$

We can carry this further. Define the following operators mapping $L_2[0,L]^q$ into $L_2[0,L]^n$:

$$Kf = g \; ; \; g(t) = B(t)^* \Phi(t)^{*-1} \int_t^L \Phi(s)^* C(s)^* f(s)\, ds$$

$$Df = g; \; g(t) = D^*(t)\, f(t) \tag{2.18}$$

Note that $K$ is a Volterra operator. Let $R$ denote the mapping $L_2[0,L]^q$ into $L_2[0,L]^q$:

$$R = (D + K)^* (D + K) \tag{2.19}$$

Then we can verify that for $f(.)$, $h(.)$ in $L_2[0,L]^q$:

$$E \left( \int_0^L [f(t),\, dY(t;\omega)] \right) \left( \int_0^L [h(t),\, dY(t;\omega)] \right) = [Rf, h] \tag{2.20}$$

where on the right side we use the innerproduct notation in $L_2[0,L]^q$. For this we have only to note that

$$\int_0^L [f(t),\, C(t)\, x(t;\omega)]\, dt$$

$$= \int_0^L [f(t),\, C(t) \int_0^t \Phi(t)\, \Phi(s)^{-1} B(s)\, dW(s;\omega)]\, dt$$

$$= \int_0^L [\Phi(t)^* C(t)^* f(t),\, \int_0^t \Phi(s)^{-1} B(s)\, dW(s;\omega)]\, dt$$

$$= \int_0^L \left[ \int_s^L \Phi(t)^* C(t)^* f(t)\, dt\, ,\; \Phi(s)^{-1} B(s)\, dW(s;\omega) \right]$$

$$= \int_0^L \left[ B(t)^* \Phi(t)^{*-1} \int_t^L \Phi(s)^* C(s)^* f(s)\, ds,\; dW(t;\omega) \right]$$

so that

$$\int_0^L [f(t),\ dY(t;\omega)]$$

$$= \int_0^L [D(t)^* f(t) + B(t)^* \Phi(t)^{*-1} \int_t^L \Phi(s)^* C(s)^* f(s)\, ds,\ \ dW(t;\omega)]$$

from which the left side of (2.20) is seen to be

$$[(D+K)f,\ \ (D+K)h]$$

<u>Problem</u> Let $J = (D+K)^*$.

Show that if $\{\phi_n(\cdot)\}$ is any complete orthonormal sequence in $L_2[0,L]^n$, we have

$$Y(t;\omega) = \sum_1^\infty \varphi_n(\omega) \int_0^t \psi_n(s)\, ds \tag{2.21}$$

where

$$\psi_n = J\phi_n,\quad J = (D+K)^*$$

and

$$\varphi_n(\omega) = \int_0^L [\phi_n(t),\ dW(t;\omega)]$$

<u>White Noise</u>

We pause briefly to explain the use of the work 'differential' in describing the system (2.14), (2.10), since they both actually involve only 'integrals'. Until recently, at least in the application oriented literature, one would write in place of (2.10):

$$\dot{x}(t;\omega) = A(t)\, x(t;\omega) + B(t)\, n(t;\omega)$$

where

$n(t;\omega)$ is defined to be 'white Gaussian noise' - a (zero-mean) Gaussian process with covariance function

$$E(n(t;\omega)\, n(s;\omega)^*) = \delta(t-s)\, I$$

where $I$ is the identity matrix, and $\delta(\cdot)$ denotes the Dirac delta function. Consistent with this definition one can calculate that

$$E\left(\left(\int_0^t n(\sigma;\omega)d\sigma\right)\left(\int_0^s n(\sigma;\omega)d\sigma\right)^*\right) = \min\,(s,t)\; I$$

and hence $n(t;\omega)$ should qualify as the 'derivative of the Wiener process'. Unfortunately, the sample paths of the Wiener process are no-where differentiable (are not of bounded variation - see e.g., Doob [2]). Hence we introduce stochastic integrals instead, and use the integral version (2.10), as the definition. In a similar manner we could have written

$$y(t;\omega) = C(t)\, x(t;\omega) + D(t)\, n(t;\omega)$$

in place of (2.14); but we use the 'correct' integral version (2.14). The difference is much more than a formalism, but cannot be appreciated until we consider non-linear operations, for example, in this volume in dealing with Radon-Nikodym derivatives, but more significantly, non-linear versions of (2.10), (2.14).

## Generalization of Stochastic Integral: Linear Case

Finally, let us note (only for the record - not needed in the sequel) that the definition of the stochastic integral can be extended to processes more general than (2.14). Let $Y(t;\omega)$, $0 \leq t \leq L$, now denote any stochastic process (not necessarily Gaussian, or continuous with probability one). Let $\phi(\cdot)$ denote a step function in $L_2(0, L)$ of the form:

$$\phi(t) = \phi(\tau_k) \qquad t_k \leq \tau_k \leq t_{k+1}$$

$$k = 0, \ldots n-1; \; t_k = 0, \quad t_n = L$$

Then define

$$\mathscr{L}(\phi) = \sum_0^{n-1} [\phi(\tau_k), (Y(t_{k+1};\omega) - Y(t_k;\omega))]$$

Suppose now that the process is such that given any two such step functions $\phi(\cdot), \Psi(\cdot)$

$$E(\mathscr{L}(\phi)\, \mathscr{L}(\Psi)) = [R\phi, \Psi] \tag{2.22}$$

where $R$ is a self-adjoint non-negative definite (bounded) linear operator mapping $L_2(0, L)$ into itself. Because the class of such step functions is dense in $L_2(0, L)$ we can see, as before, that $\mathscr{L}(\cdot)$ defines a linear bounded transformation of $L_2(0, L)$ into the Hilbert space of random variables with finite second moment. If the process is Gaussian, then the random variables $\mathscr{L}(\cdot)$ are also Gaussian. In particular, if $y(t;\omega)$ is a separable process with zero mean and continuous covariance function, and we define (Cf. p. 141)

$$Y(t;\omega) = \int_0^t y(s;\omega)ds,$$

then

$$\mathscr{L}(f) = \int_0^L [f(t), y(t;\omega)]dt$$

as we should expect, of course.

## CHAPTER III

## CONDITIONAL EXPECTATION AND MARTINGALE THEORY

In the theories of estimation and control involving stochastic differential systems, a central role is played by the concept of Martingales (due largely to Doob [1]). The Martingale theory in turn requires the notion of Conditional Expectation. In view of their importance, we shall now study these two concepts and some of the related results of particular relevance to our purposes.

Let $(\Omega, \mathscr{B}, p)$ denote a probability triple and let $\zeta$ be any (n-dimensional) random variable such that

$$E[\|\zeta\|] < \infty$$

Let $\mathscr{B}_s$ denote a sub sigma-algebra of $\mathscr{B}$. Then $\zeta$ need not be measurable $\mathscr{B}_s$. We can construct a (sort of) projection which is. Define the set function

$$\nu(B) = \int_B \zeta \, dp,$$

for every B in $\mathscr{B}_s$. $\nu(\cdot)$ is absolutely continuous with respect to the measure $p(\cdot)$ on $\mathscr{B}_s$. Hence by the Radon Nikodym theorem, there is a function $f(\omega)$, measurable $\mathscr{B}_s$, such that

$$\int_B \zeta \, dp = \int_B f(\omega) \, dp, \quad B \in \mathscr{B}_s \tag{3.1}$$

Let $\mathscr{B}_s'$ denote the sigma algebra of sets which are either in $\mathscr{B}_s$

or differ from such sets by null sets. Any function measurable with respect to $\mathscr{B}'_s$ and satisfying (3.1) will be called the conditional expectation of $\zeta$ with respect to $\mathscr{B}_s$ and denoted:

$$E[\zeta \mid \mathscr{B}_s]$$

Let us now itemize some of the properties of conditional expectations:

i) $E[\zeta \mid \mathscr{B}_s]$ is a random variable $(\Omega, \mathscr{B}'_s, p)$ and $E[\zeta \mid \mathscr{B}_s] = E[\zeta \mid \mathscr{B}'_s]$ with probability one.

ii) since $\Omega \in \mathscr{B}_s$, $\int_\Omega \zeta \, dp = \int_\Omega E[\zeta \mid \mathscr{B}_s] \, dp$

or,

$$E[\zeta] = E(E[\zeta \mid \mathscr{B}_s])$$

iii) Let $\mathscr{B}_1, \mathscr{B}_2$ be two sub sigma-algebras of $\mathscr{B}$, such that

$$\mathscr{B}_1 \subset \mathscr{B}_2$$

[This means that every set in $\mathscr{B}_1$ is also in $\mathscr{B}_2$]

Then

$$E[\zeta \mid \mathscr{B}_1] = E[E[\zeta \mid \mathscr{B}_2] \mid \mathscr{B}_1] \tag{3.2}$$

Proof

Let $f_2(\omega) = E[\zeta \mid \mathscr{B}_2]$

Let $f_1(\omega) = E[\zeta \mid \mathscr{P}_1]$

Let B be any set in $\mathscr{B}_1$. Then

$$\int_B \zeta \, dp = \int_B f_1(\omega) \, dp$$

But since B belongs to $\mathscr{B}_2$,

$$\int_B \zeta \, dp = \int_B f_2(\omega) \, dp$$

Hence for every B in $\mathscr{B}_1$,

$$\int_B f_1(\omega) \, dp = \int_B f_2(\omega) \, dp$$

or, the result follows.

iv) Let $\mathscr{B}_s$ be a sub-sigma algebra of $\mathscr{B}$ and let $h(\omega)$ be a $p \times n$ matrix valued function measurable $\mathscr{B}_s$. Suppose

$$E[\,|h(\omega)\zeta\,|\,] < \infty$$

Then

$$E[h(\omega)\zeta \mid \mathscr{B}_s] = h(\omega)\, E[\zeta \mid \mathscr{B}_s] \tag{3.3}$$

Proof

Let $h(\omega)$ be the characteristic function of a set $B_o$ in $\mathscr{B}_s$. Then for any B in $\mathscr{B}_s$

$$\int_B h(\omega)\zeta \, dp = \int_{B \cap B_o} \zeta \, dp$$

$$= \int_{B \cap B_o} f(\omega) \, dp$$

$$= \int_B h(\omega) \, f(\omega) \, dp$$

where

$$f(\omega) = E[\zeta \mid \mathscr{B}_s]$$

By linearity, this result is extended to simple functions, and by the usual limiting arguments to any function measurable $\mathcal{B}_s$

v) Jensen's Inequality

Let $f(\cdot)$ be a convex functional defined on E such that it has continuous derivatives except at a finite number of points. Let $\zeta$ be a random variable with range in E, with finite expectation. Then:

$$E(f(\zeta)/\mathcal{B}) \geq f(E(\zeta/\mathcal{B})) \text{ with pr. one.}$$

Proof Let $\nabla f(\cdot)$ denote the gradient of $f(\cdot)$. The convexity of $f(\cdot)$ implies that there exists a function $\lambda(y)$ mapping E into E such that

$$f(x) - f(y) \geq [\lambda(y), x-y]$$

At a point where the function is differentiable we have:

$$\lambda(y) = \nabla f(y)$$

and hence $\lambda(y)$ is continuous omitting a finite number of points, and hence Borel measurable. Hence

$$\lambda(E(\zeta|\mathcal{B}))$$

is also $\mathcal{B}$-measurable. Moreover with probability one

$$E((f(\zeta)-f(E(\zeta|\mathcal{B})))|\mathcal{B}) \geq E([\lambda(E(\zeta|\mathcal{B})), \zeta-E(\zeta|\mathcal{B})]|\mathcal{B}))$$

But the right side is

$$[\lambda(E(\zeta|\mathcal{B})), E(\zeta|\mathcal{B}) - E(\zeta|\mathcal{B})] = 0$$

and hence the asserted inequality follows. A useful special function for which the result applies is $f(x) = \|x\| \; (=\sqrt{[x,x]})$.

Definition: Let $\{x_\alpha(\omega)\}$ be any collection of random variables. By the smallest sigma algebra generated by this collection we shall mean the sm smallest sigma algebra containing all sets of the form:

$$[\omega | x_\alpha(\omega) \in B]$$

where B is a Borel set in E, or differs from such a set by $\omega$-sets of measure zero. We shall denote the sigma algebra by:

$$\mathscr{B}\{x_\alpha(\cdot)\}$$

Lemma (Doob): Let $x_1(\omega), \ldots x_m(\omega)$ be m random variables, and let $\mathscr{B}_m$ denote the smallest sigma algebra generated by them. Let $f(\omega)$ be measurable with respect to $\mathscr{B}_m$. Then there exists a Borel function $g(x_1, \ldots x_m)$ defined on $E^{(m)}$ such that

$$f(\omega) = g(x_1(\omega), \ldots x_m(\omega)) \text{ a.e.} \tag{3.4}$$

Proof: The main thing to note is that the smallest sigma algebra generated by the m variables is precisely the class of sets of the form:

$$[\omega | \{x_1(\omega), \ldots, x_m(\omega)\} \in U_{mn}] \ \Delta N$$

where $U_{mn}$ is a Borel set in mxn dimensions and N is any set of measure zero and $\Delta$ denotes symmetric difference. Now suppose $f(\omega)$ is a characteristic function of a set in $\mathscr{B}_m$ of the above type. Define:

$$g(x_1, \ldots x_m) = 1 \text{ on } U_{mn}$$

$$= 0 \text{ otherwise.}$$

Then $g(\ldots)$ is a Borel function and (3.4) holds. Next note that the class of functions for which (3.4) holds is a linear class. Hence (3.4) holds for simple functions $f(\cdot)$. Now given any function $f(\cdot)$ measurable $\mathscr{B}_m$, we can find a sequence of simple functions converging to it with probability one. Let $f_n(\omega)$ denote a sequence of such simple functions and let $\wedge$ denote the set of convergence. We know that

$$f_n(\omega) = g_n(x_1(\omega), \ldots x_m(\omega))$$

for Borel functions $g_n(\ldots.)$, and that $\wedge$ must be of the form:

$$[\omega \mid x_1(\omega), \ldots . x_m(\omega) \quad \varepsilon \ I] \ \Delta N$$

where I is a Borel set in mxn dimensions and N is a set of measure zero. Then $g_n(x_1, \ldots . x_m)$ converges on I. Define:

$$g(x_1, \ldots . x_m) = \text{limit } g_n(x_1, \ldots . x_m) \text{ on } I$$

$$= 0 \text{ otherwise}$$

Then $g(\ldots.)$ is a Borel function, and of course (3.4) holds. Note in particular that

$$E[\zeta \mid \mathscr{B}_m] = g(x_1(\omega), \ldots . x_m(\omega))$$

for some Borel function $g(\ldots.)$. Because of this we shall on occasion use the notation:

$$E[\zeta \mid x_1(\omega), \ldots . x_m(\omega)]$$

for the conditional expectation with respect to $\mathscr{B}_m$.

Best Mean Square Estimates

Suppose that

$$E(\|\zeta\|^2) < \infty$$

Let $f(\omega)$ be any function measurable $\mathcal{B}$ with finite second moment:

$$E(\| f(\omega) \|^2) < \infty$$

Then

$$E(\| \zeta - f(\omega)\|^2 \geq E(\| \zeta - E(\zeta/\mathcal{B})\|^2) = E(\|\zeta\|^2) - E(\|E(\zeta/\mathcal{B})\|^2) \tag{3.5}$$

The conditional expectation, in other words, is the "best mean square estimate" of $\zeta$ in terms of functions measurable $\mathcal{B}$. The actual calculation of the conditional expectation is however a difficult problem in general, except in the "Gaussian" case. See [40] and below (and Chapter VI for perhaps the most useful engineering application.)

Proof: Since by the Schwarz inequality:

$$E(|[\zeta, f(\omega)]|) \leq \sqrt{E(\|\zeta\|^2)} \cdot \sqrt{E(\|f(\omega)\|^2)} < \infty$$

and by Jensen's inequality:

$$\{E[\|E(\zeta/\mathcal{B})\|]\}^2 \leq E(\|E[\zeta/\mathcal{B}]\|^2) \leq E[\|\zeta\|^2] < \infty$$

it follows that we can use property (iv) to obtain:

$$E[\zeta - E(\zeta/\mathcal{B}), E(\zeta/\mathcal{B})] = 0 = E[\zeta - E(\zeta/\mathcal{B}), f(\omega)]$$

and hence

$$E(\|\zeta - f(\omega)\|^2) = E(\|\zeta - E(\zeta/\mathscr{B})\|^2) + E(\|E(\zeta/\mathscr{B}) - f(\omega)\|^2)$$

from which the assertion follows; and further

$$E[|\zeta - E(\zeta/\mathscr{B})|^2] = E[|\zeta|^2] + E(|E(\zeta/\mathscr{B})|^2] -2E([\zeta\, E(\zeta/\mathscr{B})])$$

$$= E[|\zeta|^2] - E(|E(\zeta/\mathscr{B})|^2)$$

Problem

Suppose in particular that: $\mathscr{B} = \mathscr{B}(x(\omega))$

Suppose further that the joint distribution of $\zeta$ and $x(\omega)$ is Gaussian, with finite second moments (matrix). Let

$$E[(\zeta - E(\zeta))(\zeta - E(\zeta))^*] = \Lambda_{\zeta\zeta}$$

$$E[(x(\omega) - E(x(\omega)))(x(\omega) - E(x(\omega))^*] = \Lambda_{xx}$$

$$E[(\zeta - E(\zeta))(x(\omega) - E(x(\omega))^*] = \Lambda_{\zeta x}$$

Then

$$E[\zeta \,|\, x(\omega)] = M(x(\omega) - E(x(\omega))) + E[\zeta]$$

where M satisfies (the discrete version of) the "Wiener-Hopf" equation:

$$\Lambda_{\zeta x} = M\Lambda_{xx}$$

Hint: It is enough to verify (3.3). If $\Lambda_{xx}$ is non-singular, then of course, M is given by:

$$M = \Lambda_{\zeta x} \Lambda_{xx}^{-1}$$

If $\Lambda_{xx}$ is singular, M is no longer unique, but one solution can be obtained as:

$$\lim_{\varepsilon \to 0} \Lambda_{\zeta x} (\Lambda_{xx} + \varepsilon I)^{-1}$$

Problem

Show that if $x(\omega)$ is a random variable such that

$$\| x(\omega) \| \leq c \leq \infty \quad \text{with probability one}$$

then so is the conditional expectation:

$$E(\zeta(\omega)/x(\omega)).$$

where of course we assume

$$E[\| \zeta(\omega) \|] < \infty$$

Independence

Let $\zeta, \eta$ denote random variables of dimension n and m respectively which are independent. Recall that this means that:

$$\text{pr.} (\zeta \in A, \eta \in B) = \text{pr.} (\zeta \in A). \ \text{pr.} (\eta \in B)$$

Let us calculate the conditional expectation:

$$E[\zeta | \eta]$$

Let B be any set in $\mathscr{B}(\eta)$. Then

$$\int_B \zeta \, dp = \int_\Omega f(\omega) \, \zeta \, dp$$

where

$$f(\omega) = \text{Identity matrix on } B, \text{ and zero otherwise.}$$

Then $f(\omega)$ and $\zeta$ are independent, and hence we have that:

$$\int_\Omega f(\omega) \, \zeta \, dp = E(\zeta) \, p(B)$$

Hence

$$\int_B \zeta \, dp = \int_B E(\zeta) \, dp$$

Or,

$$E(\zeta | \eta) = E(\zeta) \text{ with probability one}$$

Martingales-Discrete Parameter

A sequence of random variables $\{\zeta_n\}$ such that:

$$E[|\zeta_n|] < \infty$$

is called a Martingale if

$$E[\zeta_n | \mathscr{B}_{n-1}] = \zeta_{n-1}$$

where $\mathscr{B}_n$ is the smallest sigma algebra generated by the first n variables. Note in particular that:

$$E[\zeta_n | \zeta_{n-1}] = E[E[\zeta_n | \mathscr{B}_{n-1}] | \zeta_{n-1}]$$

$$= \zeta_{n-1}$$

Suppose $\{y_n\}$ is a sequence of independent, random variables with finite expectation, but with zero mean. Then

$$\zeta_n = \sum_1^n y_k$$

is a Martingale.

Moreover we have the following generalization of the Kolmogorov inequality to Martingales: (Doob):

Lemma: Let $\{x_n\}$ be a Martingale. Then:

$$pr.[\operatorname*{Max}_{j \leq n} |x_j(\omega)| \geq \epsilon] \leq (1/\epsilon^2) \; E[|x_n|^2]$$

Proof:

Let

$$A_k = [\omega | \; |x_k(\omega)| \geq \epsilon]$$

$$B_k = A_k - \bigcup_{j=1}^{k-1} A_j$$

Define

$$z_k(\omega) = x_k(\omega) \text{ for } \omega \text{ in } B_k$$

$$= 0 \text{ otherwise}$$

Then $z_k$ is measurable $\mathscr{B}(x_1, \ldots x_k)$, and hence:

$$E([z_k, x_n]) = E(E([z_k, x_n] \mid x_k)) = E([z_k, E(x_n | x_k)])$$

$$= E([z_k, x_k])$$

where the Martingale property is invoked in the last equality. We can now readily proceed as in the proof of the original Kolmogorov inequality:

$$E(|x_n|^2) \geq \sum_1^n \int_{B_k} |x_n|^2 \, dp = \sum_1^n \int_{B_k} \left\{ |x_n - x_k|^2 + |x_k|^2 + 2[x_k, x_n - x_k] \right\} dp$$

and

$$\int_{B_k} [x_k, x_n - x_k] \, dp = E([z_k, x_n - x_k]) = E[z_k, x_k - x_k] = 0$$

so that

$$E(|x_n|^2) \geq \sum_1^n \int_{B_k} (|x_k|^2) \, dp \geq (\epsilon^2) \, \text{pr.}(\bigcup_k B_k) = (\epsilon^2) \, \text{pr.}(\bigcup_k A_k)$$

which yields the inequality sought.

Doob's Martingale Convergence Theorem:

The following special case of Doob's theorem will be useful to us in the sequel:

Theorem 3.1 Suppose $\{\zeta_n\}$ is a Martingale such that

$$E(|\zeta_n|) \leq c < \infty$$

Then the sequence $\zeta_n$ converges with probability one to a variable $\zeta$ such that

$$E(|\zeta|) \leq c$$

Proof See Doob ([ 2 ] p. 319)

As an example of how this result is used, let us consider the following canonical situation. Let $x(\omega)$ be a variable with finite expectation, and let $y_n(\omega)$ be a sequence of random variables, and let $\mathscr{B}_n$ denote the sigma algebra generated by the first n variables $y_1, \ldots y_n$. Let $\mathscr{B}_\infty$ denote the sigma algebra generated by the whole sequence. Then we can state:

Corollary

$$\zeta_n = E(x(\omega)/\mathscr{B}_n)$$

defines a martingale sequence which converges with probability one, to:

$$\zeta = E(x(\omega)/\mathscr{B}_\infty)$$

and

$$E(|\zeta_n - \zeta|) \to 0 \tag{3.6}$$

Proof From

$$E(\zeta_n/\mathscr{B}_{n-1}) = E(E(x(\omega)/\mathscr{B}_n)/\mathscr{B}_{n-1}) = E(x(\omega)/\mathscr{B}_{n-1}) = \zeta_{n-1}$$

we can see that $\zeta_n$ is a martingale. Next

$$E(|\zeta_n|) = E(|E(\zeta_{n+1}/\zeta_n)|)$$

and Jensen's inequality yields:

$$E(|\zeta_n|) \leq E(|\zeta_{n+1}|) \leq E(|x(\omega)|).$$

Hence $\zeta_n$ converges with probability one. Let $\hat{\zeta}$ denote the limit. Assume first that $|x(\omega)| \leq c$ with pr. 1. Then it is readily seen that $\zeta_n$ has the same property for every n. Hence for any set B in $\mathscr{B}_n$, we observe that

$$\begin{aligned} \int_B \zeta_n \, dp &= \int_B \zeta_m dp \ , \quad m \geq n \\ &= \lim_m \int_B \zeta_m \, dp = \int_B \hat{\zeta} \, dp \end{aligned}$$

and hence

$$E(\hat{\zeta}|\mathcal{B}_n) = \zeta_n$$

Hence

$$E(\zeta - \hat{\zeta}|\mathcal{B}_n) = 0 \text{ for every } n$$

Or, really

$$\int_B (\zeta-\hat{\zeta})dp = 0 \text{ for every } B \text{ in } \mathcal{B}_n \text{ for every } n. \quad (3.7)$$

But the class of sets B such that B belongs to $\mathcal{B}_n$ for some n is a field, and it generates $\mathcal{B}_\infty$. Hence (3.7) holds for every B in $\mathcal{B}_\infty$, or,

$$\zeta = \hat{\zeta} \text{ with probability one}$$

since both variables are measurable with respect to $\mathcal{B}_\infty$. Let us next consider the general case. Let $c > 0$ and let

$$x(\omega)' = x(\omega) \text{ on } [\omega|\ |x(\omega)| \leq c]; \text{ zero otherwise}$$

and let

$$x(\omega)'' = x(\omega) \text{ on the complementary set:}$$

$$[\omega|\ |x(\omega)| > c]; \text{ and zero otherwise}$$

Then letting

$$\zeta'_n = E[x(\omega)'|\mathcal{B}_n] \ ; \quad \zeta' = E[x(\omega)'|\mathcal{B}_\infty]$$

$$\zeta''_n = E[x(\omega)''|\mathcal{B}_n] \ ; \quad \zeta'' = E[x(\omega)''|\mathcal{B}_\infty]$$

we note that both $\zeta'_n$ and $\zeta''_n$ converge with probability one.

Denoting the limits by $\hat{\zeta}'$ and $\hat{\zeta}''$ respectively, we have that

$$\hat{\zeta}' = \zeta' \text{ with pr. one}$$

Hence for any B in $\mathscr{B}_n$,

$$\left|\int_B (\zeta-\hat{\zeta})dp\right| = \left|\int_B (\zeta''-\hat{\zeta}'')dp\right| \leq \int_\Omega |\zeta''-\hat{\zeta}''|\,dp$$

Now

$$\int_\Omega |\zeta''-\hat{\zeta}''|\,dp \leq E(|\zeta''|) + E[|\hat{\zeta}''|]$$

$$\leq 2 \int_\Omega |x(\omega)''|\,dp$$

and hence goes to zero as $c \to \infty$. Hence

$$\int_B (\zeta-\hat{\zeta})dp = 0$$

and hence for every B in $\mathscr{B}_\infty$ and hence (3.6) follows.

A special case of interest (in connection with best mean square estimation) is:

$$E(|\zeta|^2) < \infty$$

In that case

$$E[|\zeta_n|^2] \leq E(|\zeta_{n+1}|^2) \leq E(|\zeta|^2)$$

Hence

$$E[|\zeta_n|^2]$$

converges and since for $n > m$

$$E[|\zeta_n - \zeta_m|^2] = E[|\zeta_n|^2] - E[|\zeta_m|^2]$$

it follows $\zeta_n$ converges in the mean of order two. Since it converges with probability one to $\zeta$, we have that

$$\lim_n E[|\zeta - \zeta_n|^2] = 0$$

$$E[|\zeta|^2] = \lim_n E[|\zeta_n|^2]$$

Continuous Parameter Martingales:

Let us now turn to continuous parameter Martingales which form the central part of our study. Let $Z(t;\omega)$ be a stochastic process, $t \in T$ where $T$ is assumed to be an interval of the real line. Let $\mathscr{F}(t)$ be a sigma algebra of measurable sets such that $Z(t;\omega)$ is measurable $\mathscr{F}(t)$ and let

$$\mathscr{F}(t_1) \subset \mathscr{F}(t_2) \text{ for } t_1 < t_2$$

Then $Z(t;\omega)$ is said to be a Martingale with respect to $\mathscr{F}(\cdot)$, or simply a Martingale if

(i) $E(|Z(t;\omega)|) < \infty, \quad t \in T$

(ii) $E(Z(t;\omega) \mid \mathscr{F}(s)) = Z(s;\omega), \; s < t$

Let $\mathscr{B}(t)$ denote the sigma algebra generated by $Z(s;\omega)$ for $s \leq t$. Then it is clear that $Z(t;\omega)$ continues to be a Martingale with respect to $\mathscr{B}(t)$.

The Wiener process $W(t;\omega)$ is a Martingale with respect to $\mathcal{B}(t)$. In fact:

$$E(W(t;\omega) - W(s;\omega) \Big| W(\sigma_1;\omega), \ldots . W(\sigma_n;\omega)) = 0$$

for any finite number of indices $\sigma_1, \ldots, \sigma_n, \leq s$. The class of sets B such that

$$B \in \mathcal{B}(W(\sigma_1;\omega), \ldots . W(\sigma_n;\omega)) , \quad \sigma_i \leq s$$

for some finite number of indices, forms a field which generates $\mathcal{B}(s)$, and hence

$$E(W(t;\omega) \,|\mathcal{B}(s)) = W(s;\omega) \text{ for } s \leq t.$$

Actually more is true because non-overlapping increments are actually independent:

$$E[W(t;\omega)-W(s;\omega))(W(t;\omega)-W(s;\omega))^* \,|\, \mathcal{B}(s)] = (t-s)\, I_n, \quad s < t$$

where $I_n$ is the indentity matrix. Next let $F(s)$ be any rectangular m-by-n matrix function, Lebesgue measurable. For any rectangular matrix A, we define:

$$\|A\|^2 = \text{Trace } AA^* = \text{Trace } A^*A$$

Assume now that

$$\int_T \|F(s)\|^2 \, ds < \infty$$

Then

$$Z(t;\omega) = \int_0^t F(s)\, dW(s;\omega) \quad , \quad t \in T$$

is also a Martingale with respect to $\mathscr{B}(t)$. In fact:

$$Z(t;\omega) - Z(s;\omega) = \int_s^t F(\sigma)\, dW(\sigma;\omega)$$

and for any simple function $f(\cdot)$, we know that

$$E\left[\int_s^t f(\sigma)\, dW(\sigma;\omega) \,|\mathscr{B}(s)\right] = 0$$

and hence since $F(.)$ is the limit of such simple functions,

$$E\left[\int_s^t F(\sigma)\, dW(\sigma;\omega) \,|\mathscr{B}(s)\right] = 0$$

Moreover we know that

$$E[(Z(t;\omega) - Z(s;\omega))(Z(t;\omega) - Z(s;\omega))^*) \,|\mathscr{B}(s)] =$$

$$\int_s^t F(\sigma) . F(\sigma)^* d\sigma$$

We shall now generalize the stochastic integral (stil in the linear set-up) to Martingales, specifically to a class of Martingales which, following Nelson [ 8 ], we shall call $R_2$ Martingales. Since we shall only be concerned with such Martingales in the sequel, we shall not insert the qualification unless another kind is involved. By an $R_2$ Martingale, we shall mean a Martingale with the additional property that:

$$E[[(Z(t;\omega) - Z(s;\omega))(Z(t;\omega) - Z(s;\omega))^* \,|\mathscr{F}(s)] = \int_s^t P(\sigma)\, d\sigma$$

where P(s) is a non-negative definite matrix function, Lebesgue measurable, and

$$\text{Tr.} \int_T P(s)\, ds < \infty$$

To define the stochastic integral we follow the same procedure as before. We wish to define:

$$\int_T f(t)\, dZ(t;\omega)$$

for a class or rectangular (m-by-n say to be specific) matrix functions. The class will be the Hilbert space of Lebesgue measurable functions with inner-product defined by:

$$[f,g] = \int_T [f(s),\ g(s)P(s)]\ ds \quad ; \text{(here } [a,b] = \text{Tr. } ab^*\text{)}.$$

It will be convenient to keep using the generic notation $\mathscr{H}$ to denote the space. For most purposes we may assume that P(s) is continuous; in any case we shall always assume that we can avoid the trivial case where it is zero almost everywhere. To define the integral we begin with simple functions which we know are dense in $\mathscr{H}$. For simple functions it is readily verified that

$$E([\int_T f(t)dZ(t;\omega)\ ,\ \int_T g(t)dZ(t;\omega)]) = [f,g]$$

and the mapping being linear into $L_2(\Omega)$ of m-dimensional random variables, we can complete the definition as before. It should be noted that the basic Martingale property implies that

$$E[(Z(t_3;\omega) - Z(t_2;\omega))(Z(t_2;\omega) - Z(t_1;\omega)^*)\ |\mathscr{B}(t_2)] = 0,\ t_1 < t_2 < t_3$$

An Example:

Let $W(t;\omega)$ denote the Wiener process, and let $F(s)$ denote an m-by-n matrix function, Lebesgue measurable such that

$$F(s)F(s)^* > 0 \quad \text{a.e.}$$

and

$$\int_T \| F(s) \|^2 \, ds < \infty$$

where we shall now take $T = [0, 1]$.

Define

$$Z(t;\omega) = \int_0^t F(s) \, dW(s;\omega)$$

Then $Z(t;\omega)$ is an $R_2$ Martingale. Let

$$a(s) = \sqrt{F(s)F(s)^*}$$

which is then non-singular a.e. Let

$$b(s) = a(s)^{-1}$$

Then the function $b(s)$ has the property

$$\int_0^1 [b(s), b(s) \, F(s)F(s)^*] \, ds = m < \infty$$

Hence we can define

$$K(t;\omega) = \int_0^t b(s) \, dZ(s;\omega), \; 0 \leq t \leq 1$$

Thus defined, we obtain a Wiener process in m dimensions. To verify this, we note that for any two $m$ dimensional vectors ($m$-by-1 matrices) $u$, $v$, we have that:

$$E([(u^*\int_s^t b(\sigma)dZ(\sigma;\omega))(v^*\int_s^t b(\sigma)\,dZ(\sigma;\omega))] =$$

$$= \int_s^t [u^*b(\sigma),\ v^*b(\sigma)F(\sigma)F(\sigma)^*]d\sigma$$

$$= u^*\int_s^t b(\sigma)F(\sigma)F(\sigma)^*\, b(\sigma)^*v\; d\sigma$$

so that for $s < t$

$$E[(K(t;\omega) - K(s;\omega))(K(t;\omega) - K(s;\omega))^*] = \int_s^t b(\sigma)F(\sigma)F(\sigma)^*\, b(\sigma)^*\, d\sigma$$

$$= (t-s)\, I_m$$

where $I_m$ is the $m$-by-$m$ unit matrix. Of course $K(t;\omega)$ is Gaussian.

Moreover, we have the representation (cf. Doob, Nelson):

$$Z(t;\omega) = \int_0^t a(s)\, dK(s;\omega)$$

For this we have only to note that the partial sums (we assume that $a(t)$ is continuous from now on):

$$\sum_0^{n-1} a(t_i)(K(t_{i+1};\omega) - K(t_i;\omega)), \qquad 0 \le t_i < t_{i+1} \le t;$$

$$0 = t_0;\ t_n = t$$

$$= \sum_{0}^{n-1} a(t_i) \int_{t_i}^{t_{i+1}} b(s)\, dZ(s;\omega)$$

$$= \sum_{0}^{n-1} a(t_i)(b(t_i)(Z(t_{i+1};\omega)-Z(t_i;\omega))) - \sum_{0}^{n-1} a(t_i) \int_{t_i}^{t_{i+1}} (b(s)-b(t_i))dZ(s;\omega)$$

The first term

$$= Z(t;\omega)$$

while the second term goes to zero as the maximal subdivision length goes to zero. For

$$E( \sum_{0}^{n-1} a(t_i) \int_{t_i}^{t_{i+1}} (b(s)-b(t_i))dZ(s;\omega))^2 =$$

$$\sum_{0}^{n-1} \int_{t_i}^{t_{i+1}} \| a(t_i)(b(s) - b(t_i))F(s) \|^2 \, ds$$

and

$$\| a(t_i)b(s)-b(t_i))F(s) \|^2 = \| a(t_i)(b(s) - b(t_i))\, a(s) \|^2$$

$$= \| a(t_i) - a(s) \|^2$$

and $a(s)$ is uniformly continuous on $[0, 1]$.

Finally, we note that by using Doob's Martingale inequality, we can obtain

$$\int_0^t F(s)\, dZ(s;\omega)$$

as a continuous function of $t$. for almost all $\omega$.

# CHAPTER IV

## Radon-Nikodym Derivatives with Respect to Wiener Measure

Let $W(t;\omega)$ denote the n-dimensional Wiener process, $0 \leq t \leq 1$, inducing the Wiener measure on $\mathscr{C} = C(0,1)$ as we have indicated. Let $\varphi(\omega)$ denote a (Borel) measurable function mapping $\mathscr{C}$ into $\mathscr{C}$. Then $\varphi(\cdot)$ induces a measure on the Borel sets of $\mathscr{C}$ given by:

$$p_\varphi(B) = p_W(\varphi^{-1}(B))$$

where $p_W$ denotes the Wiener measure. For example, often $\varphi(\cdot)$ may be defined by means of a stochastic integral:

$$x(t;\omega) = L(t) \int_0^t M(s) dW(s;\omega)$$

where say the functions $L(\cdot)$, $M(\cdot)$ are continuous. Of particular importance is to determine when the induced measure is absolutely continuous with respect to Wiener measure, and then to evaluate the Radon-Nikodym derivative[+]. In the present volume we shall mostly be concerned with the case where $\varphi(\cdot)$ is a linear transformation (or affine transformation).

The following general result plays a fundamental role in our approach.

Theorem 4.1

Suppose $\varphi(\cdot)$ is a measurable map of $\mathscr{C}$ into $\mathscr{C}$. Let $p_\varphi(\cdot)$ denote

---

[+]See Chapter VIII for an application to System Identification problems.

the induced measure. Let $\{\phi_n\}$ be any complete orthonormal system in $L_2(0,1)^{(n)}$. Let

$$\zeta_n = \int_0^1 [\phi_n(t), dW(t;\omega)]$$

Let $\mathscr{B}_n$ denote the sigma algebra $\mathscr{B}(\zeta_1, \ldots\ldots, \zeta_n)$. Suppose for every B in $\mathscr{B}_n$, we have

$$p_\varphi(B) = \int_B H_n(\omega)\, dp_W$$

In other words, $p_\varphi$ is absolutely continuous with respect to $p_W$ on $\mathscr{B}_n$, for every n, and $H_n(\omega)$ denotes the corresponding derivative. Suppose for some $\alpha$, $0 < \alpha < 1$, we have:

$$\sup_n E[H_n(\omega)^{1+\alpha}] < \infty$$

Then $p_\varphi$ is absolutely continuous with respect to $p_W$, and the derivative is given by

$$H(\omega) = \lim H_n(\omega)$$

omitting a set of Wiener measure zero

Proof We observe first of all that the sequence of random variables $H_n(\omega)$ is a Martingale. For, for any set B in $\mathscr{B}_n$, we have that

$$p_\varphi(B) = \int_B H_{n+1}(\omega)\, dp_W = \int_B H_n(\omega)\, dp_W$$

since $\mathscr{B}_n$ is increasing with n. Also:

$$E(H_n(\omega)) = 1$$

Hence by the Doob Martingale convergence theorem, we know that $H_n(\omega)$

converges with probability one. Let us denote the limit function by $H(\omega)$. Now the condition

$$\sup_n E\,(H_n(\omega)^{1+\alpha}) < \infty\,;\quad 0 < \alpha < 1$$

implies "uniform integrability" of the sequence $H_n(.)$; that is to say:

$$\sup_n \int_{E_n(a)} H_n(\omega)\, dp_W$$

where

$$E_n(a) = [\omega \mid H_n(\omega) > a]$$

goes to zero as "a" goes to infinity. This in turn (see any standard text, Doob [2], or Neveu [24] p. 52) implies that

$$E(\,|H_n(\omega) - H(\omega)|\,)$$

goes to zero with n, and in particular then, for every set B in $\mathscr{B}_n$, we have:

$$p_\varphi(B) = \int_B H_n(\omega)\, dp_W = \lim_m \int_B H_m(\omega)\, dp_W = \int_B H(\omega)\, dp_W$$

Hence denoting by $\mathscr{B}_\infty$ the smallest sigma-algebra containing all sets in $\mathscr{B}_n$ for every n, we have:

it follows that

$$p_\varphi(B) = \int_B H(\omega)\, dp_W,\quad B \in \mathscr{B}_\infty$$

It is clearly enough to show that $\mathscr{B}_\infty$ differs from $\mathscr{B}_{\mathscr{C}}$, the sigma-algebra of Borel sets by sets of Wiener measure zero. For this let $f(\cdot)$ be any element in $\mathscr{H} = L_2(0,1)^{(n)}$. Let

$$\zeta(\omega) = \int_0^1 [f(t),\ dW(t;\omega)]$$

Then it is enough to show that $\zeta(\omega)$ is measurable $\mathcal{B}_\infty$. Let

$$\xi_k(\omega) = \sum_1^k [f, \phi_m] \zeta_m(\omega)$$

Then of course $\xi_m(\omega)$ is measurable $\mathcal{B}_\infty$ and

$$E((\zeta(\omega) - \xi_m(\omega))^2) \longrightarrow 0$$

Let

$$\hat{\zeta}(\omega) = E(\zeta(\omega) | \mathcal{B}_\infty)$$

Then

$$\hat{\zeta}(\omega) - \xi_m(\omega) = E((\zeta(\omega) - \xi_m(\omega)) | \mathcal{B}_\infty)$$

and hence

$$E((\hat{\zeta}(\omega) - \xi_m(\omega))^2) \leq E((\zeta(\omega) - \xi_m(\omega))^2) \longrightarrow 0$$

Hence

$$E(|\zeta(\omega) - \xi_m(\omega)|^2) = 0$$

and hence

$$E[\zeta(\omega) | \mathcal{B}_\infty] = \zeta(\omega)$$ omitting a set of Wiener measure zero

Or,

$\zeta(\omega)$ is measurable $\mathcal{B}_\infty'$ as we started to show.

<u>Corollary</u> Suppose $\varphi(\omega)$ is a linear transformation such that for

any $f(\cdot)$ in $\mathscr{H}$, and any linear Borel set U,

$$\varphi(\cdot)^{-1}\left[\omega \,\Big|\, \int_0^1 [f(t),\ dW(t;\omega)] \in U\right] = \left[\omega \,\Big|\, \int_0^1 [h(t),\ dW(t;\omega)]\right] \in U$$

where

$$h = Mf;$$

and M is a linear bounded transformation mapping $\mathscr{H}$ into itself.

Suppose

$$M^* M = (I + J)^{-1}$$

where J is Hilbert- Schmidt. Then $p_\varphi$ is absolutely continuous with respect to Wiener measure. Denoting the derivative by $H(\omega)$, we have further that:

$$E(H(\omega)^{1+\alpha}) < \infty, \text{ for some } \alpha,\ 0 < \alpha < 1$$

Proof: By definition, J is self-adjoint and Hilbert-Schmidt. Hence let $\{\phi_n\}$ denote the orthonormalized sequence of eigen-functions, including those corresponding to zero eigen values, so that $\{\phi_n\}$ is a complete orthonormal base. Let $\gamma_n$ denote the corresponding eigen values. Then, we must have:

$$(1 + \gamma_n) > 0 \ ; \quad \sum_1^\infty \gamma_n^2 < \infty$$

Let

$$\zeta_n(\omega) = \int_0^1 [\phi_n(t),\ dW(t;\omega)]$$

and let

$$\mathscr{B}_n = \mathscr{B}(\zeta_1, \ldots, \zeta_n)$$

Define

$$H_n(\omega) = \left(\prod_1^n \sqrt{(1 + \gamma_i)}\right) \cdot \exp - (1/2) \sum_1^n \gamma_i \zeta_i(\omega)^2 \tag{4.1}$$

Then it is not difficult to verify that for any B in $\mathscr{B}_n$.

$$p_\varphi(B) = \int_B H_n(\omega)\, dp_W$$

For it is enough to verify this for sets of the form:

$$B = [\omega \mid \zeta_i(\omega) \in U_i, \quad i = 1, \ldots n, \ U_i \text{ linear Borel sets}]$$

Now

$$p_\varphi(B) = p_W(\varphi^{-1}B) = p_W(\omega \mid \int_0^1 [h_i(t), dW(t;\omega)] \in U_i, \ i = 1, \ldots n)$$

$$= \text{probability } \eta_i(\omega) \in U_i, \ i = 1, \ldots n$$

where

$$\eta_i(\omega) = \int_0^1 [h_i(t), dW(t;\omega)]$$

are independent Gaussians of zero means and variance

$$[h_i, h_i] = [M\phi_i, M\phi_i] = [M^*M\phi_i, \phi_i] = (1+\gamma_i)^{-1}$$

and hence

$$p_\varphi(B) = \int_{U_1} \cdots \int_{U_n} \prod_1^n \sqrt{(1+\gamma_i)} \left(\frac{1}{\sqrt{2\pi}}\right)^n \exp - (1/2) \sum_1^n x_i^2(1+\gamma_i).\ dx_1 \ldots dx_n$$

$$= \int_{U_1} \cdots \int_{U_n} \left(\prod_1^n \sqrt{(1+\gamma_i)}\right) \left(\exp(-1/2) \sum_1^n \gamma_i x_i^2\right) \frac{1}{\sqrt{2\pi}^{\,n}} \exp(-1/2) \sum_1^n x_i^2.\ dx_1 \ldots dx_n$$

$$= \int_B H_n(\omega)\, dp_W$$

Next let

$$\gamma = \min_i \gamma_i$$

Then we can clearly find a number $\alpha$, $0 < \alpha < 1$ such that

$$(1 + (1+\alpha)\gamma) > 0$$

Then we can calculate that : for every i,

$$E( \exp (-1/2) ( 1 + \alpha ) \gamma_i \zeta_i (\omega)^2) = ( (1 + (1+\alpha) \gamma_i)^{-1/2}$$

and hence that:

$$E( H_n(\omega)^{1+\alpha}) = \prod_1^n \sqrt{\frac{(1+\gamma_i)^{1+\alpha}}{(1 + (1+\alpha) \gamma_i)}}$$

We shall show that the right side is bounded, and calculate a bound. For this, let us take the logarithm of the right side. Then

$$\operatorname{Log} E(H_n(\omega)^{1+\alpha} ) = \sum_1^n (1/2)(1+\alpha) \operatorname{Log} (1+\gamma_i) - \sum_1^n (1/2) \operatorname{Log}(1+(1+\alpha)\gamma_i)$$

$$= \sum_1^n (1/2) (1+\alpha) \int_0^{\gamma_i} ((1+x)^{-1} - (1+x +\alpha x)^{-1}) \, dx$$

Now

$$\left| \int_0^{\gamma_i} ((1+x)^{-1} - (1+x+\alpha x)^{-1}) dx \right| \leq \alpha \gamma_i^2/(2)(1-|\gamma|)(1-(1+\alpha)|\gamma| )$$

so that, using $\|\cdot\|^2_{H.S.}$ to denote Hilbert-Schmidt norm:

$$E(H_n(\omega)^{1+\alpha}) \leq \exp k(\alpha) \sum_1^\infty \gamma_i^2 = \exp k(\alpha) \|J\|^2_{H.S}$$

where

$$k(\alpha) = (1/4) \alpha (1+\alpha) (1-|\gamma|)^{-1} (1 - (1+\alpha)|\gamma|)^{-1}$$

Hence we can apply the theorem. Thus $H_n(\omega)$ converges in the mean of order one to $H(\omega)$ which is the Radon-Nikodym derivative sought. By Fatou's Lemma, we have

$$E(H(\omega)^{1+\alpha} ) \leq \exp k(\alpha) \|J\|^2_{H.S.} \tag{4.2}$$

We can obtain a "closed form" expression for $H(\omega)$ in the following way.

$$|\operatorname{Log}(1+\gamma_i) - \gamma_i| = \left| \int_0^{\gamma_i} (-x/(1+x))dx \right| \leq \gamma_i^2 /(1-|\gamma|)$$

so that the 'infinite product'

$$\prod_1^{\infty} (1+\gamma_i) \exp - \gamma_i$$

is convergent (to a finite non-zero limit). Denote it by $\mu$. Note also that

$$S_n = \sum_1^n \gamma_i (\zeta_i(\omega)^2 - 1)$$

converges with probability one being the sum of independent variates with

$$E(S_n^2) = \sum_1^n \gamma_i^2 (3 + 1 - 2)$$

and the series on the right of course converges. Hence

$$H_n(\omega) = \prod_1^n \left(\sqrt{1+\gamma_i} \exp - \gamma_i/2\right) \exp - (1/2) \sum_1^n \gamma_i (\zeta_i(\omega)^2 - 1)$$

$$\longrightarrow \sqrt{\mu} \exp(-1/2) \sum_1^{\infty} \gamma_i (\zeta_i(\omega)^2 - 1) = H(\omega) \ldots \quad (4.3)$$

Note that in particular this implies that

$$E(\exp(-1/2) \sum_1^{\infty} \gamma_i (\zeta_i(\omega)^2 - 1)) = \frac{1}{\sqrt{\mu}} \leq \exp + \frac{\|J\|^2_{H.S.}}{2(1+\gamma)}$$

Remark: If $J$ is trace-class (as in many cases that follow), then

$$\sum_1^{\infty} |\gamma_n| < \infty$$

so that

$$\sum_1^{\infty} \log(1+\gamma_n) \text{ and } \sum_1^{\infty} \gamma_n \zeta_n(\omega)^2$$

converges and hence (4.1) converges to (the R.N. derivative):

$$H(\omega) = \left(\prod_1^\infty \sqrt{1+\gamma_i}\right) \exp - 1/2 \sum_1^\infty \gamma_i \zeta_i (\omega)^2 \tag{4.2a}$$

Example: Let $x(t;\omega)$ denote the continuous solution of:

$$x(t;\omega) = \int_0^t A(t)\, x(t;\omega) dt + W(t;\omega) \quad 0 \leq t \leq 1 \tag{4.3a}$$

where let us assume that $A(t)$ is continuous in $[0,1]$. Then the measure induced by the process $x(t;\omega)$ on $C$, is absolutely continuous with respect to Wiener measure. In this case we can explicitly write the solution as:

$$x(t;\omega) = \Phi(t) \int_0^t \Phi(s)^{-1} dW(s;\omega)$$

which then defines the mapping $\varphi(\cdot)$ explicitly. This shows that the mapping $M$ is given by

$$M = I + K$$

where

$$K f = g; \quad g(t) = \Phi(t)^{*-1} \int_t^1 \Phi(s)^* A(s)^* f(s) ds$$

Since $K$ is thus a Volterra operator, we note that $(I+K)$ has a bounded inverse. In fact a simple calculation shows that

$$(I + K^*)^{-1} = (I - L)$$

where $L$ is again a Volterra operator defined by:

$$Lf = g \,; \quad g(t) = A(t) \int_0^t f(s) ds$$

Hence we have

$$(M^*M)^{-1} = (I-L^*)(I-L) \tag{4.4}$$

Thus

$$J = L^*L - (L + L^*) \tag{4.5}$$

and is obviously Hilbert-Schmidt.

It is pertinent to ask where (4.3) can be expressed as a functional on the (Wiener) process without having to go through the eigenfunctions. Before we do this in general, it is interesting to consider the special case where

$$A(t) = A(t)^* = A$$

Then the main simplification is that $(L + L^*)$ is trace-class.[+] Indeed

$$(L + L^*) f = g; \quad g(t) = A \int_0^1 f(s)ds$$

so that the operator has finite-dimensional range, and

$$\mathrm{Tr}(L + L^*) = \mathrm{Tr}\ A$$

Hence $J$ is trace-class, since $L^*L$ being the product of Hilbert-Schmidt operators is clearly trace-class. The implication of this is that now,

$$\sum_1^\infty |\gamma_n| < \infty \tag{4.6}$$

+ In fact it should be noted that in case $A(t)$ is independent of t, $(L+L^*)$ is trace-class if and only if $A=A(t)$ is self-adjoint. See Appendix I for relevant facts concerning Volterra operators.

so that the infinite product

$$\prod_{i=1}^{\infty} (1+\gamma_i)$$

converges. Indeed we can actually evaluate this product by the general formula that

$$\sum_1^{\infty} \text{Log } (1 + \gamma_i) = \text{Tr } (\text{Log}(I-L^*)(I-L)) = -\text{Tr } (L+L^*) \tag{4.7}$$

which is valid whenever $(L + L^*)$ is trace-class. A proof is given in Appendix I. Since (4.6) holds,

$$-\frac{1}{2} \sum_1^{\infty} \gamma_k \zeta_k(\omega)^2 \tag{4.8}$$

converges also, with probability one. We can go on to evaluate the limit. For this let us note that

$$\gamma_k \zeta_k(\omega) = \sum_1^{\infty} [J\phi_k, \phi_m] \zeta_m(\omega)$$

Since

$$[J\phi_k, \phi_m] = [L\phi_k, L\phi_m] - [\phi_k, (L + L^*) \phi_m]$$

and

$$[\phi_k, (L+L^*)\phi_m] = [\int_0^1 \phi_k(t)dt, A \int_0^1 \phi_m(t)dt]$$

$$\sum_1^{\infty} A \int_0^1 \phi_m(t)dt \;\; \zeta_m(\omega) = A\, W(1;\omega)$$

and letting

$$L\phi_k = \Psi_k$$

we have:

$$\sum_{1}^{\infty} \Psi_k(t)\, \zeta_k(\omega) = A\, W(t;\omega)$$

Hence we obtain finally that

$$\gamma_k \zeta_k(\omega) = \int_0^1 [\Psi_k(t),\ A\, W(t;\omega)]\, dt - [\int_0^1 \phi_k(t)dt,\ AW(1;\omega)]$$

Multiplying by $\zeta_k(\omega)$ and summing on $k$, the second term clearly yields

$$-[W(1;\omega),\ AW(1;\omega)]$$

while the first term

$$\sum_{1}^{\infty} \int_0^1 [\zeta_k(\omega)\, \Psi_k(t),\ A\, W(t;\omega)]\, dt$$

$$= \int_0^1 [A\, W(t;\omega),\ A\, W(t;\omega)]\, dt$$

For this we only need to note that

$$\int_0^1 [A\, W(t;\omega) - \sum_{1}^{n} \zeta_k(\omega) \Psi_k(t),\ A\, W(t;\omega)]\, dt$$

$$= \int_0^1 [W(t;\omega) - \sum_{1}^{n} \zeta_k(\omega) \int_0^t \phi_k(s) ds\ ,\ \ A^*A\, W(t;\omega)]\, dt$$

and

$$\left\{ E(| \int_0^1 [W(t;\omega) - \sum_{1}^{n} \zeta_k(\omega) \int_0^t \phi_k(s) ds\ ,\ \ A^*A\, W(t;\omega)]\, dt\ |) \right\}^2$$

$$\leq \int_0^1 E(\| W(t;\omega) - \sum_{1}^{n} \zeta_k(\omega) \int_0^t \phi_k(s) ds \|^2 dt) \int_0^1 E(\| A^*AW(t;\omega) \|^2) dt$$

(cf. Problem p. 21)

which goes to zero. Hence finally

$$\sum_{1}^{\infty} -\frac{1}{2} \gamma_n \zeta_n(\omega)^2 = -\frac{1}{2}\int_0^1 [AW(t;\omega), AW(t;\omega)]\, dt + \frac{1}{2}[AW(1;\omega), W(1;\omega)]$$

Hence we have:

$$H(\omega) = \exp\left(-\frac{1}{2} \operatorname{Tr} A - \frac{1}{2}\int_0^1 [AW(t;\omega), AW(t;\omega)]\, dt\right.$$

$$\left. + \frac{1}{2}[AW(1;\omega), W(1;\omega)]\right) \tag{4.9}$$

So far we have been concerned with mapping $\mathscr{C}$ into $\mathscr{C}$. This is of course not essential. Thus let $Y(t;\omega)$ be any stochastic process with continuous sample functions for almost all $\omega$. Then for $0 \leq t \leq 1$,

$$\varphi(\omega) = Y(\cdot;\omega)$$

is a measurable map into $\mathscr{C}$, measureable with respect to Borel sets in $\mathscr{C}$. Hence we can define an induced measure $p_\varphi$ as before. Of particular interest is the case where $Y(t;\omega)$ is a Gaussian process and the stochastical integral

$$\int_0^1 f(t)\, dY(t;\omega)$$

can be defined. We can then state the following theorem.

Theorem 4.2 Let $Y(t;\omega)$ be a Gaussian process with continuous sample paths. Let the process dimension be $n$. Suppose the stochastic integral can be defined for every $f(\cdot)$ in $L_2(0,1)^{(n)}$ such that

$$E\left(\int_0^1 [f(t),\ dY(t;\omega)]\right)\left(\int_0^1 [g(t),\ dY(t;\omega)]\right) = [Rf, g] \ \ldots \qquad (4.10)$$

where $R$ is a linear transformation of $L_2(0,1)^{(n)}$ into itself, and has the form:

$$R = (I + J)^{-1}$$

where $J$ is self-adjoint Hilbert-Schmidt operator. Then the measure induced by the process is absolutely continuous with respect to Wiener measure.

<u>Proof</u> Let $c$ denote the generic element in $\mathscr{C}$, and let $W(t;c)$ denote the Wiener process,

$$W(t;c) = c(t)$$

We need only note that for any $f(\cdot)$ in $L_2(0,1)^{(n)}$, and any linear Borel set $U$,

$$\varphi(\cdot)^{-1}\left([c \,|\, \int_0^1 [f(t),\ dW(t;c)] \in U\right) = [\omega \,|\, \int_0^1 [f(t), dY(t;\omega)] \in U]$$

The proof can clearly proceed exactly as in Theorem 2.3, using (4.10).

<u>Example</u> Let us consider the process $Y(t;\omega)$ defined by (2.16), and take the case where $L = 1$ and

$$D(t)D(t)^* > 0 \qquad \text{a.e.}$$

and assume that

$$b(t) = \sqrt{((D(t)D(t)^*)^{-1})}$$

is essentially bounded. Let

$$\tilde{Y}(t;\omega) = \int_0^t b(s)\, dY(s;\omega)$$

Then the measure induced by the $\tilde{Y}(\cdot;\omega)$ process is absolutely continuous with respect to Wiener measure. Indeed

$$\int_0^1 [f(t), d\tilde{Y}(t;\omega)] = \int_0^1 [b(t)f(t), dY(t;\omega)]$$

so that, introducing the operator

$$b\, f = g; \quad g(t) = b(t)\, f(t)$$

we have that

$$E\left(\left(\int_0^1 [f(t), d\tilde{Y}(t;\omega)]\right)\left(\int_0^1 [h(t), d\tilde{Y}(t;\omega)]\right)\right) = [Rf, h]$$

where

$$R = b^*(D+K)^* (D + K)\, b = I + b^* K^* K b + b^* D^* K b + b^* K^* D b \tag{4.11}$$

Now because $(R - I)$ is compact and self-adjoint, it is clear that $R$ has a bounded inverse unless for some non-zero $f(\cdot)$

$$[Rf, f] = 0$$

Or, equivalently,

$$[(D + K)b\, f,\ (D + K)\, b\, f] = 0$$

or,

$$(D + K)\, b\, f = 0$$

Or, with slight abuse of notation:

$$D(t)^{*}\, b(t)\, f(t) + K\, b\, f = 0$$

But, multiplying by $D(t)$, and noting that

$$(D(t)D(t)^{*})^{-1} = b(t)\, b(t)$$

we obtain that

$$b\, f + (b(t)^{2}\, D(t))\, K\, b\, f = 0$$

But this implies that $(b\, f)$ must be zero, since the second term involves a Volterra operator. Since $b(t)$ is non-singular, $f$ must be zero. Hence $Rf$ is zero only if $f$ is zero, or $R$ has a bounded inverse. Further it is clear that

$$R^{-1} = I + J$$

where $J$ is self-adjoint, and Hilbert- Schmidt.

As a final example of Radon-Nikodym derivatives, let us consider the case of a non-zero mean. Thus let

$$Y(t;\omega) = W(t;\omega) + \int_0^t m(s)ds \tag{4.1}$$

where $m(\cdot)$ is in $L_2[0, 1]^{(n)}$. Then the measure induced by the process $Y(t;\omega)$ is absolutely continuous with respect to Wiener process.

For this, let $\{\phi_n\}$ be a complete orthonormal system and it is readily seen that for B in $\mathscr{B}(\zeta_1, .., \zeta_n)$, where, as usual:

$$\zeta_n = \int_0^1 [\phi_n(t), dW(t;\omega)]$$

we have, denoting the new measure by $p_Y$,

$$p_Y(B) = \int_B H_n(\omega)\, dp_W$$

where

$$H_n(\omega) = \exp -\frac{1}{2}\left(\sum_1^n [m, \phi_k]^2 - \sum_1^n 2\, \zeta_k(\omega)\, [m, \phi_k]\right)$$

from which it follows that the Radon-Nikodym derivative is given by:

$$H(\omega) = \exp -\frac{1}{2}\left(\int_0^1 |m(t)|^2\, dt - 2\int_0^1 [m(t), dW(t;\omega)]\right) \tag{4.13}$$

## CHAPTER V

### The Ito Integral

— . —

In this chapter we come to one of the main tools in the theory of stochastic differential systems - the stochastic integral named after its inventor K. Ito. The development we follow is that of Doob [2]. After discussing what is perhaps its main feature that distinguishes it from 'ordinary' integrals, we show how it can be used to obtain a 'closed-form' expression for the Radon-Nikodym derivatives. What is more, it enables us to obtain closed forms even in cases where the approach of Chapter IV fails ((4.2a) is <u>not</u> valid).

### The Ito Integral: Definition and Basic Properties

The Ito integral can be looked upon as the non-linear generalization of the stochastic integrals we have so far considered. The main extension is that the integrand will now be a random process also. Let $Z(t;\omega)$ denote an $R_2$ Martingale, with $\mathscr{F}(t)$ the increasing sigma-algebra and

$$E((Z(t;\omega) - Z(s;\omega))(Z(t;\omega) - Z(s;\omega))^*)|\mathscr{F}(s)) = \int_s^t P(\sigma)d\sigma \qquad (5.$$

We wish to define the integral

$$\int_0^1 f(t;\omega)\, dZ(t;\omega)$$

where $f(t;\omega)$ is an m-by-n matrix valued random process with the following properties:

(i) $f(t;\omega)$ is measurable jointly in $t$ and $\omega$, in $t$ with respect to Lebesgue measure; $f(t;\omega)$ is measurable $\mathcal{F}(t)$ for each fixed $t$ (a.e. with respect to Lebesgue measure). The significance of this is that $f(t;\omega)$ depends only on the 'past' of the process $Z(t;\omega)$. Sometimes this is indicated by saying that $f(t;\omega)$ is 'non-anticipatory', or 'physically realizable'.

(ii) $$\int_0^1 E([f(t;\omega),\ f(t;\omega)P(t)])\ dt < \infty \tag{5.2}$$

Let $\mathcal{H}$ denote the class of such functions; it is clearly a linear space.

Introduce an inner-product in $\mathcal{H}$ by:

$$[f, g] = \int_0^1 E[f(t;\omega),\ g(t;\omega)P(t)]\ dt \tag{5.3}$$

(Here let us recall that in the integrand:

$$[a, b] = \text{Tr. } ab^* = \text{Tr. } a^*b)$$

By a "simple" function in $\mathcal{H}$ we shall mean a function of the form:

$$f(t;\omega) = v_i(\omega)\ , \quad t_i \le t < t_{i+1}\ , \quad 0 \le i \le n-1,$$

$$t_0 = 0\ , \quad t_n = 1$$

It is implicit of course that $v_i(\omega)$ is measurable $(t_i)$. As in the linear case, we define the integral for such a simple function by:

$$\sum_{i=0}^{n-1} v_i(Z(t_{i+1};\omega) - Z(t_i;\omega))$$

Note that

$$E([v_i(Z(t_{i+1};\omega) - Z(t_i;\omega)),\ v_j(Z(t_{j+1};\omega) - Z(t_j;\omega))])$$

$$= E(E(\text{ditto} \mid F(t_{j+1})) \text{ if } t_{j+1} \le t_i, \text{ and hence zero.}$$

$$= \text{Tr. } E(v_i(Z(t_{i+1};\omega) - Z(t_i;\omega))(Z(t_{i+1};\omega) - Z(t_i;\omega))^* v_i^* =$$

$$\int_{t_i}^{t_{i+1}} E[v_i, v_i P(s)]\, ds, \quad \text{for } i = j$$

Hence by a similar calculation as in the linear case we see that for any two simple functions, $f(t;\omega)$, $g(t;\omega)$ in $\mathscr{H}$, we have:

$$E([\int_0^1 f(t;\omega)\, dZ(t;\omega),\ \int_0^1 g(t;\omega)\, dZ(t;\omega)]) = [f, g]$$

In other words we have an inner-product preserving linear transformation from the class of simple functions in $\mathscr{H}$, into the space $L_2(\Omega)$ of matrix valued (m-by-n) $\omega$-random variables with finite second moment.

Next we wish to show that $\mathscr{H}$ is a Hilbert space and that the simple functions as above are actually dense in it. That $\mathscr{H}$ is a Hilbert space is immediate. For, the class of functions $f(t;\omega)$ jointly measurable in $t$ and $\omega$, and such that

$$\int_0^1 E\| f(t;\omega)\sqrt{P(t)}\|^2\, dt < \infty$$

is of course a Hilbert space, and $\mathscr{H}$ is a subspace of it. If $f_n(t;\omega)$ is a Cauchy sequence in $\mathscr{H}$, then there is a subsequence converging pointwise almost everywhere to the limit function $f(t;\omega)$ and hence the limit function has the property that $f(t;\omega)$ is measurable $\mathscr{F}(t)$ a.e. in $t$, and hence is in $\mathscr{H}$. To show that the simple functions are dense in is a little more involved. First we observe that the class of essentially bounded function are dense in $\mathscr{H}$. Let $f(t;\omega)$ denote such a function with bound $M$. Define for each integer $n$:

$$f_n(t;\omega) = n \int_{t-1/n}^{t} f(s;\omega)\, ds = n \int_0^{1/n} f(t-s;\omega) ds$$

Then $f_n(\cdot;\cdot)$ is in $\mathscr{H}$. Moreover

$$\| f(\cdot) - f_n(\cdot) \|^2$$

using Schwartz inequality is

$$\leq E(n \int_0^{1/n} ds \int_0^1 \| (f(t;\omega) - f(t-s;\omega)) \sqrt{P(t)} \|^2 dt$$

Now for each $\omega$, omitting a set of measure zero, invoking a known property of $L_2$-spaces:

$$h(t;\omega) = \int_0^1 \| (f(t;\omega) - f(t-s;\omega)) \sqrt{P(t)} \|^2 dt$$

$$\longrightarrow 0 \quad \text{as} \quad s \longrightarrow 0$$

and

$$h(s;\omega)| \leq 4M^2 \int_0^1 \text{Tr. } P(t)\, dt$$

Hence

$$\| f(\cdot) - f_n(\cdot) \| \longrightarrow 0$$

But for each $n$, and $0 < \Delta < 1/n$:

$$f_n(t+\Delta;\omega) - f_n(t;\omega) = n \int_0^{\Delta} (f(t+s;\omega)ds + f(t-1/n+s;\omega))ds$$

so that

$$E(\|f_n(t+\Delta;\omega) - f_n(t;\omega)\|^2)$$

$$< 2\, n^2 \Delta\, E(\int_0^{\Delta} \|f(t+s;\omega)\|^2\, ds + \int_0^{\Delta} \|f(t-1/n+s;\omega\|^2\, ds)$$

using Schwartz inequality, and hence

$$< 2\, n^2(\Delta)\,(\text{const.}) \longrightarrow 0 \quad \text{with} \quad \Delta$$

Hence it follows that $f_n(t;\omega)$ is uniformly continuous in $t$ 'in the mean square sense'. In other words we have shown that such functions are dense in $\mathscr{H}$. On the other hand, if $g(t;\omega)$ in $\mathscr{H}$, and is uniformly continuous in the mean square, define the sequence of simple functions:

$$g_n(t;\omega) = g(k/2^n;\omega), \quad k/2^n \leq t < (k+1)/2^n; \; 0 \leq k \leq 2^n - 1$$

so that for given $\varepsilon > 0$,

$$E(\|g(t;\omega) - g_n(t;\omega)\|^2) \quad < \varepsilon \qquad \text{for all } n \text{ sufficiently large.}$$

$$\|g_n - g\|^2 \leq \int_0^1 E[\|g(t;\omega) - g_n(t;\omega)\|^2]\; \text{Tr. } P(t)\, dt$$

$$\longrightarrow 0$$

Hence the simple functions are dense in $\mathscr{H}$.

Here is an elementary canonical example which illustrates the difference between the ordinary integral and the Ito integral. Let $W(t;\omega)$ be the Wiener process. Then the Ito integral

$$\int_0^1 [W(t;\omega), \; dW(t;\omega)]$$

is clearly definable. Let us attempt what is essentially an integration by parts. For this we begin with an approximating finite sum:

$$\sum_{i=0}^{m-1} [W(t_i;\omega), W(t_{i+1};\omega) - W(t_i;\omega)], \; t_0 = 0, \ldots . t_i < t_{i+1}, \; t_m = 1$$

which we can rewrite as:

$$= - \sum_0^{m-1} [W(t_{i+1};\omega) - W(t_i;\omega) \;, \; W(t_{i+1}, \omega) - W(t_i;\omega)]$$

$$+ \sum_0^{m-1} [W(t_{i+1};\omega) \;, W(t_{i+1};\omega) - W(t_i;\omega)]$$

Here the second term can be expressed:

$$\sum_{i=0}^{m-1} \sum_{j=0}^{i} [W(t_{j+1};\omega) - W(t_j;\omega), \; W(t_{i+1};\omega) - W(t_i;\omega)]$$

and by interchanging the order of summation, we have that this is the same as:

$$\sum_{j=0}^{m-1} \sum_{i=j}^{m-1} [W(t_{j+1};\omega) - W(t_j;\omega) \;, \; W(t_{i+1};\omega) - W(t_i;\omega)]$$

$$= \sum_{j=0}^{m-1} [W(t_{j+1};\omega) - W(t_j;\omega) \;, \; W(1;\omega) - W(t_j;\omega)]$$

$$= \sum_{j=0}^{m-1} [W(t_{j+1};\omega) - W(t_j;\omega), \; W(1;\omega)] - \sum_{j=0}^{m-1} [W(t_{j+1};\omega) - W(t_j;\omega), \; W(t_j;\omega)]$$

and since the first term in this

$$= [W(1;\omega), \; W(1;\omega)]$$

and the second term is the sum we originally started with, we have

$$\sum_{i=0}^{m-1} [W(t_i;\omega), W(t_{i+1};\omega) - W(t_i;\omega)] = \frac{1}{2}[W(1;\omega), W(1;\omega)]$$

$$- \frac{1}{2} \sum_{0}^{m-1} \| W(t_{i+1};\omega) - W(t_i;\omega) \|^2$$

Now since the sum on the left converges as the subdivision size shrinks, the second term on the right converges, in the mean square sense also. But

$$E(\| W(t_{i+1};\omega) - W(t_i;\omega) \|^2) = n(t_{i+1} - t_i)$$

and hence

$$E\left( \sum_{0}^{m-1} (\| W(t_{i+1};\omega) - W(t_i;\omega) \|^2 - n(t_{i+1} - t_i))^2 \right.$$

$$= E \sum_{0}^{m-1} \| W(t_{i+1};\omega) - W(t_i;\omega) \|^4 - n^2 \sum_{0}^{m-1} (t_{i+1} - t_i)^2$$

$$= 0 (\max |t_{i+1} - t_i|) \longrightarrow 0$$

Hence

$$\int_0^1 [W(t;\omega), dW(t;\omega)] = +\frac{1}{2}[W(1;\omega), W(1;\omega)] - \frac{1}{2}(\mathrm{Tr.}\, I)$$

The significant point is the appearance of the second term, which is a characteristic feature of the Ito integral.

Problem: Show that if we define the partial sum slightly differently:

$$\sum_{i=0}^{m-1} [W((t_i + t_{i+1})/2;\omega), \quad W(t_{i+1};\omega) - W(t_i;\omega)]$$

we get a completely different answer in the limit. In fact the difference

$$\sum_{i=0}^{m-1} [W((t_i+t_{i+1})/2;\omega) - W(t_i;\omega), W(t_{i+1};\omega) - W(t_i;\omega)]$$

$$\longrightarrow \frac{1}{2}(\mathrm{Tr.}\ I).$$

More generally, taking

$$\tau_i = \gamma(t_{i+1}-t_i) \qquad 0 < \gamma < 1$$

show that the limit of the partial sums:

$$\sum_{i=0}^{m-1} [W(t_i+\tau_i;\omega) - W(t_i;\omega), W(t_{i+1};\omega) - W(t_i;\omega)]$$

$$\longrightarrow (\gamma)(\mathrm{Tr.}\ I)$$

Hence the partial sums:

$$\sum_{0}^{m-1} [W(t_i+\tau_i;\omega), W(t_{i+1};\omega) - W(t_i;\omega)]$$

converge to

$$\frac{1}{2}[W(1;\omega), W(1;\omega)] + (\gamma - \frac{1}{2})(\mathrm{Tr.}\ I).$$

Let us now prove a useful generalization of this result. Let $W(t;\omega)$ denote the Wiener process, and let $H(t;\omega)$ be an n-by-1 process defined by:

$$H(t;\omega) = \int_0^t L(s)\, dW(s;\omega) \qquad 0 \le t \le 1$$

where $L(\cdot)$ is continuous. Let us calculate the Ito integral:

$$\int_0^1 [H(t;\omega),\ dW(t;\omega)]$$

Let us begin with an approximating sum:

$$\sum_{i=0}^{n-1} [H(t_i;\omega),\ (W(t_{i+1};\omega) - W(t_i;\omega))]$$

$$= (-1) \sum_{i=0}^{n-1} [(H(t_{i+1};\omega) - H(t_i;\omega)),\ ((W(t_{i+1};\omega) - W(t_i;\omega)))]$$

$$+ \sum_{i=0}^{n-1} [H(t_{i+1};\omega),\ (W(t_{i+1};\omega) - W(t_i;\omega))]$$

As before,

$$\sum_{i=0}^{n-1} [H(t_{i+1};\omega),\ (W(t_{i+1};\omega) - W(t_i;\omega))] = \sum_{j=0}^{n-1} [(H(t_{j+1};\omega)$$

$$- H(t_j;\omega)),\ (W(1;\omega) - W(t_j;\omega))] = [H(1;\omega),\ W(1;\omega)]$$

$$- \sum_{j=0}^{n-1} [(H(t_{j+1};\omega) - H(t_j;\omega)),\ W(t_j;\omega)]$$

Transposing, and taking limits, and noticing that

$$\sum_{j=0}^{n-1} [(H(t_{j+1};\omega) - H(t_j;\omega)),\ W(t_j;\omega)]$$

defines in the limit the Ito integral:

$$\int_0^1 [dH(t;\omega),\ W(t;\omega)]$$

we have:

$$\int_0^1 [H(t;\omega),\ dW(t;\omega)] + \int_0^1 [W(t;\omega),\ dH(t;\omega)]$$

$$= [H(1;\omega), W(1;\omega)] - \lim \sum_0^{n-1} [H(t_{i+1};\omega) - H(t_i;\omega)), (W(t_{i+1};\omega) - W(t_i;\omega))]$$

As before the limit exists in the mean square sense, and we shall now show that the limit is actually equal to:

$$\int_0^1 \mathrm{Tr.}(L(t))\, dt$$

But this follows readily from the fact that:

$$E\Big(\sum_0^{n-1} [H(t_{i+1};\omega) - H(t_i;\omega)),\ W(t_{i+1};\omega) - W(t_i;\omega)]\Big)$$

$$= \sum_0^{n-1} E([\int_{t_i}^{t_{i+1}} L(s) dW(s;\omega),\ \int_{t_i}^{t_{i+1}} dW(s;\omega)])$$

$$= \int_0^1 \mathrm{Tr.}\ L(t)\, dt$$

Hence

$$E\Big(\Big(\sum_0^{n-1} \Big\{[\int_{t_i}^{t_{i+1}} L(s)\, dW(s;\omega),\ \int_{t_i}^{t_{i+1}} dW(s;\omega)] - \int_{t_i}^{t_{i+1}} \mathrm{Tr.}\ L(t)dt\Big\}\Big)^2\Big) \quad (5.4)$$

$$= \sum_0^{n-1} E([\int_{t_i}^{t_{i+1}} L(s)dW(s;\omega),\ \int_{t_i}^{t_{i+1}} dW(s;\omega)] - \int_{t_i}^{t_{i+1}} \mathrm{Tr.}\ L(s)ds)^2$$

$$= \sum_0^{n-1} E([\int_{t_i}^{t_{i+1}} L(s)dW(s;\omega),\ \int_{t_i}^{t_{i+1}} dW(s;\omega)])^2 - \sum_0^{n-1} \Big(\int_{t_i}^{t_{i+1}} \mathrm{Tr.}L(s)ds\Big)^2$$

and

$$E\left(\left[\int_{t_i}^{t_{i+1}} L(s)dW(s;\omega), \int_{t_i}^{t_{i+1}} dW(s;\omega)\right]^2\right)$$

$$\leq E\left(\left\| \int_{t_i}^{t_{i+1}} L(s)dW(s;\omega)\right\|^2\right) n(t_{i+1}-t_i)$$

$$\leq \left(\int_{t_i}^{t_{i+1}} \text{Tr. } L(s)L(s)^* ds\right) n(t_{i+1}-t_i)$$

so that (5.4) clearly goes to zero as maximal length of subdivision goes to zero. Hence we finally have the result:

$$\int_0^1 \left[\int_0^t L(s)dW(s;\omega), \ dW(t;\omega)\right] = \left[\int_0^1 L(s)dW(s;\omega), \int_0^1 dW(s;\omega)\right]$$

$$- \int_0^1\left[\int_0^t dW(s;\omega), \ L(t)dW(t;\omega)\right] - \int_0^1 \text{Tr. } L(t)\, dt \ldots\ldots \tag{5.5}$$

and of course again, the unusual thing to note is the appearance of the third term on the right.

This can clearly be generalized into a more symmetric form:

$$\int_0^1\left[\int_0^t L(s)dW(s;\omega), \ M(t)dW(t;\omega)\right] = \left[\int_0^1 L(s)dW(s;\omega), \int_0^1 M(s)dW(s;\omega)\right]$$

$$-\int_0^1 \left[L(t)dW(t;\omega), \int_0^t M(s)dW(s;\omega)\right]$$

$$-\int_0^1 \text{Tr. } L(t)M(t)^*\, dt \tag{5.6}$$

Note that the right side can be put in the form:

$$\int_0^1 \left[M(t)^* \int_0^t L(s)dW(s;\omega), \ dW(t;\omega)\right] = \int_0^1 \left[dW(t;\omega), \ L(t)^* \int_t^1 M(s)\, dW(s;\omega)\right]$$

$$- \text{Tr. } \int_0^1 L(t)\, M(t)^*\, dt \qquad (5.$$

but the integral (first term) on the right has to be interpreted just in the way we got it by adding the first terms on the right in (5.6).

R-N Derivatives using Ito Integrals

We shall now see how to express the Radon-Nikodym derivatives in terms of Ito integrals. First of all, let us consider the special result (4.9). Using (5.6), we have (by setting $L(t) = A$; $M(t) = $ Identity; and $A = A^*$):

$$\int_0^1 [AW(t;\omega), dW(t;\omega)] = \frac{1}{2}[AW(1;\omega), W(1;\omega)] - \frac{1}{2}\int_0^1 \text{Tr. } A \, dt$$

and substituting into (4.9) we have:

$$H(\omega) = \exp -\frac{1}{2}\int_0^1 [AW(t;\omega), AW(t;\omega)]\,dt + \int_0^1 [AW(t;\omega), dW(t;\omega)] \tag{5.8}$$

and we have the advantage that the Trace term disappears. This expression for the Radon-Nikodym derivative is actually valid for the general case of (4.4). But first let us consider the case where $J$ in Theorem 4.2 is trace-class. We begin with a theorem of interest in itself.

Theorem 5.1 Let $L$ denote a Volterra operator, mapping $L_2(0,1)^{(n)}$ into itself.

$$Lf = g; \quad g(t) = \int_0^t L(t;s)\, f(s)ds; \quad L(t;s) \text{ continuous in } 0 \leq s \leq t \leq 1$$

where $L$ is also trace-class. Then for any complete orthonormal system $\{\phi_k\}$, we have:

$$\int_0^1 [\int_0^t L(t;s)dW(s;\omega), dW(t;\omega)] = \sum_1^\infty \sum_1^\infty \zeta_k \zeta_m [L\phi_k, \phi_m] \tag{5.9}$$

where

$$\zeta_k = \int_0^1 [\phi_k(t), \; dW(t;\omega)]$$

and the convergence of the infinite series is at least in the mean of order two.

Proof First let us consider the convergence of the series.

Let

$$\varphi_{n,m} = \sum_{k=1}^{n} \sum_{j=1}^{m} [L\phi_k, \phi_j] \, \zeta_k \, \zeta_j$$

Then consider the remainder term:

$$E((\varphi_{n,m} - \varphi_{n+p,\,m+p})^2) = E((x_1+x_2+x_3)^2)$$

where

$$x_1 = \sum_{j=1}^{m} \sum_{k=n+1}^{n+p} a_{kj} \, \zeta_j \, \zeta_k$$

$$x_2 = \sum_{k=n+1}^{n+p} \sum_{j=m+1}^{m+p} a_{kj} \, \zeta_j \, \zeta_k$$

$$x_3 = \sum_{k=1}^{n} \sum_{j=m+1}^{m+p} a_{kj} \, \zeta_j \, \zeta_k$$

where

$$a_{kj} = [L\phi_k, \phi_j]$$

For any finite sum:

$$E((\Sigma_j \, \Sigma_k \; a_{jk} \, \zeta_j \, \zeta_k)^2) \leq 3 \sum_j a_{jj}^2 + \sum_{k \neq j} \sum (a_{jk}^2 + a_{kj}^2) + \sum_{k \neq j} \sum a_{kk} \, a_{jj}$$

$$= 2 \sum_k \sum_j a_{jk}^2 + (\sum_j a_{jj})^2$$

Now since $L$ is trace class,

$$\sum_1^\infty |a_{jj}| \text{ converges.}$$

Again

$$\Sigma_k \Sigma_j \; [L\phi_j, \phi_k]^2 \leq \Sigma_j \|L\phi_j\|^2$$

and of course $L$ being Hilbert-Schmidt,

$$\sum_{j=1}^\infty \|L\phi_j\|^2 < \infty$$

These estimates applied to each of the terms $x_1, x_2, x_3$, shows that the remainder term can be made arbitrarily small for all $p$, for $n, m$ sufficiently large. Moreover

$$E[(\sum_1^\infty \sum_1^\infty \zeta_k \zeta_j [L\phi_k, \phi_j])^2] \leq 2 \sum_1^\infty \|L\phi_k\|^2 + (\sum_1^\infty [L\phi_k, \phi_k])^2$$
$$= 2 \sum_1^\infty \|L\phi_k\|^2 \tag{5.10}$$

using

$$E(\sum_1^\infty \sum_1^\infty \zeta_k \zeta_m [L\phi_k, \phi_m]) = \sum_1^\infty [L\phi_k, \phi_k] = 0$$

since $L$ being Volterra, its trace must be zero. Let us next consider the Ito integral on the left of (5.9). For an approximating finite sum:

$$\eta_n = \sum_{i=0}^{n-1} \left[ \int_0^{t_i} L(t_i;s)\, dW(s;\omega),\; W(t_{i+1};\omega) - W(t_i;\omega) \right],$$

$$t_i < t_{i+1};\; t_0 = 0;\; t_n = 1,$$

we have, by using the expansion:

$$\eta_n = \sum_{m=1}^\infty \left( \sum_{k=1}^\infty \zeta_k \zeta_m \sum_{i=0}^{n-1} \left[ \int_0^{t_i} L(t_i;s)\, \phi_k(s) ds,\; \int_{t_i}^{t_{i+1}} \phi_m(s) ds \right] \right) \tag{5.11}$$

Define the operator $L_n$ by

$$L_n f = g \; ; \; g(t) = \int_0^{t_i} L(t_i;s)f(s)ds \, , \quad t_i \leq t < t_{i+1}$$

Then $L_n$ is clearly finite dimensional, and hence trace-class. But the sum in (5.11) can be written

$$\sum_m \sum_k \zeta_k \zeta_m [L_n\phi_k, \phi_m] = \eta_n$$

Let

$$\eta = \sum_1^\infty \sum_1^\infty \zeta_k \zeta_m [L\phi_k, \phi_m]$$

Then since $L_n, L$ are Volterra operators of trace-class:

$$E(\eta_n - \eta) = \text{Tr.}\,(L_n - L) = 0$$

and more important

$$E[(\eta_n - \eta)^2] = E(\sum_1^\infty \sum_1^\infty \zeta_k \zeta_m [(L_n - L)\phi_k, \phi_m])^2$$

$$= 0(\|L_n - L\|^2_{H.S.}) \tag{5.12}$$

using (5.10).

But the Hilbert-Schmidt norm of $(L_n - L)$ clearly goes to zero, and hence (5.12) goes to zero.

Since we only really needed the fact that

$$\sum_1^\infty |a_{jj}| \text{ converges}$$

we can state:

Corollary: Suppose L is not necessarily trace-class but (L+L)* is trace-class. Then

$$\int_0^1 [\int_0^t L(t;s)\, dW(s;\omega),\ dW(t;\omega)] = \sum_1^\infty \sum_1^\infty \zeta_k \zeta_m [L\phi_k, \phi_m]$$

$$- 1/2\ \mathrm{Tr.}(L+L^*)$$

REMARK: The condition that L(t;s) be continuous is not necessary; square-integrability is enough. For, we can always find a sequence $L_n$ with continuous kernels approximating L in the Hilbert-Schmidt norm and then use (5.12).

Theorem 5.2 Let $Y(t;\omega)$, $0 \le t \le 1$, be a Gaussian process with zero mean and continuous sample paths such that the stochastic integral

$$\int_0^1 [f(t),\ dY(t;\omega)]$$

can be defined for each $f(\cdot)$ in $L_2(0,1)^n$ where

$$E((\int_0^1 [f(t), dY(t;\omega)])\ (\int_0^1 [g(t), dY(t;\omega)])) = [Rf, g]$$

for $f(\cdot)$, $g(\cdot)$ in $L_2(0,1)^n$, where

$$R = (I + J)^{-1}$$



$$J = L^*L - (L + L^*)$$

and L has the form:

$$L f = h\ ;\ h(t) = M(t) \int_0^t L(s)\, f(s)\, ds$$

$M(\cdot)$, $L(\cdot)$ being continuous square matrices. Further it is assumed that J is trace-class. Then the measure induced on $\mathscr{C}$ by the mapping:

$$\varphi(\omega) = Y(\cdot;\omega)$$

is absolutely continuous with respect to Wiener measure, and the derivative is given by:

$$H(v) = \exp\left\{-\frac{1}{2}\int_0^1 [W_1(t;v), W_1(t;v)]\,dt + \int_0^1 [W_1(t;v), dW(t;v)]\right\} \tag{5.13}$$

where

$$W_1(t;v) = M(t)\int_0^t L(s)dW(s;v)$$

and W(t;v) is the Wiener process with

$$W(t;v) = v(t)\ ,\ v(\cdot) \in \mathscr{C}$$

<u>Proof:</u> Let $\{\phi_k\}$ denote the complete orthonormal system of eigenvectors of J with corresponding eigenvalues $\gamma_k$. Then as we know from Theorem 4.2 and Cor., the Radon-Nikodym derivative is given by:

$$H(v) = (\prod_k \sqrt{(1+\gamma_k)})\,(\exp -\frac{1}{2}\sum_1^\infty \gamma_k \zeta_k^2) \tag{5.13a}$$

where

$$\zeta_k = \int_0^1 [\phi_k(t), dW(t;v)]$$

Also if J is trace-class, so is (L+L*) and from (4.7):

$$\Pi\sqrt{(1+\gamma_k)} = \exp -\frac{1}{2}\ \text{Tr.}\ (L + L^*)$$

Now we can write:

$$\sum_1^\infty \gamma_k \zeta_k^2 = \sum_1^\infty \zeta_k \int_0^1 [\Psi_k(t), \ dW(t;v)]$$

where

$$J \phi_k = \Psi_k$$

and

$$\int_0^1 [\Psi_k(t), dW(t;v)] = \sum_1^\infty [J \phi_k, \phi_m] \zeta_m$$

$$= \sum_1^\infty [L\phi_k, L\phi_m] \zeta_m - \sum_1^\infty [(L+L^*) \phi_k, \phi_m] \zeta_m$$

It is readily seen that

$$\sum_{m=1}^\infty [L\phi_k, L\phi_m] \zeta_m = \int_0^1 [L\phi_k , W_1(t;v)]dt$$

while writing:

$$L + L^* = Q + P$$

where

$$Qf = g; \quad g(t) = M(t) \int_0^t L(s) f(s) ds - L(t)^* \int_0^t M(s)^* f(s)ds$$

$$Pf = g; \quad g(t) = L(t)^* \int_0^1 M(s)^* f(s)ds$$

we note that $P$ is finite dimensional, and hence trace-class, and hence $Q$ must be trace-class, and $Q$ being Volterra, its trace must be zero. Hence we obtain:

$$\text{Tr.}\ (L + L^*) = \text{Tr.}\ P = \int_0^1 \text{Tr.}\ L(t)\,M(t)\,dt = \int_0^1 \text{Tr.}\ M(t)L(t)dt \qquad \ldots (5.14)$$

We thus obtain, Q being now Volterra and trace-class,

$$\sum_1^\infty \gamma_k \zeta_k^2 = \int_0^1 [W_1(t;v),\ W_1(t;v)]\,dt - \sum\sum \zeta_k \zeta_m [Q\phi_k, \phi_m]$$

$$- \left[\int_0^1 M(s)^* dW(s;v),\ \int_0^1 L(s)dW(s;v)\right]$$

and using Theorem 5.1 we have that

$$\sum\sum \zeta_k \zeta_m [Q\phi_k, \phi_m] = \int_0^1 [M(t)\int_0^t L(s)dW(s;v),\ dW(t;v)]$$

$$- \int_0^1 [L(t)^* \int_0^t M(s)^* dW(s;v), dW(t;v)]$$

But from (5.6) we have that

$$\int_0^t [L(t)^* \int_0^t M(s)^* dW(s;v),\ dW(t;v)]$$

$$= \left[\int_0^1 M(s)^*\, dW(s;v),\ \int_0^1 L(s)dW(s;v)\right] - \int_0^1 \left[M(t)\int_0^t L(s)dW(s;v), dW(t;v)\right]$$

$$- \text{Tr.} \int_0^1 M(t)^*\ L(t)^*\ dt$$

Hence, substituting, we finally obtain:

$$\sum_1^\infty \gamma_n \zeta_k^2 = \int_0^1 [W_1(t;v),\ W_1(t;v)]\,dt - 2\int_0^1 [M(t)\int_0^t L(s)dW(s;v), dW(t;v)]$$

$$- \int_0^1 \text{Tr.}\ M(t)^*\ L(t)^* dt$$

from which (5.13) readily follows, upon using (5.14).

We shall state as a Corollary the slight generalization to the case where there is a non-zero mean.

Corollary: Suppose

$$Y_1(t;\omega) = Y(t;\omega) + \int_0^t m(s)ds$$

where

$$\int_0^1 \|m(t)\|^2 \, dt < \infty ,$$

Then the measure induced by the process $Y_1(t;\omega)$ is absolutely continuous with respect to Wiener measure, and the corresponding Radon-Nikodym derivative is given by

$$H_1(v) = \exp -1/2 \left\{ \int_0^1 [\hat{m}(t) - m(t) - h(t), \hat{m}(t) - m(t) - h(t)]\, dt \right.$$

$$\left. +2 \int_0^1 [\hat{m}(t) - m(t) - h(t), \; dW(t;v)] \right\} \tag{5.15}$$

where

$$h(t) = M(t) \int_0^t L(s)\, dW(s;v)$$

$$\hat{m}(t) = M(t) \int_0^t L(s)\, m(s)ds$$

**Proof :** With the same notation as in the proof of the Theorem, let

$$m_k = \int_0^1 [\phi_k(t), \; m(t)]dt$$

**Then we have:**

$$H_1(\omega) = \prod_k \sqrt{(1+\gamma_k)} \quad \exp - 1/2 \sum_1^{\infty} \left\{ (1+\gamma_k)(\zeta_k - m_k)^2 - \zeta_k^2 \right\}$$

**Noting that**

$$\sum_1^{\infty} \gamma_k(\zeta_k - m_k)^2 = \int_0^1 [h(t) - \hat{m}(t),\ h(t) - \hat{m}(t)]\, dt$$

$$- 2\int_0^1 [h(t) - \hat{m}(t),\ dW(t;v)]$$

$$+ 2\int_0^1 [h(t) - \hat{m}(t),\ m(t)]\, dt$$

$$-2 \sum_1^{\infty} \zeta_k m_k = -2\int_0^1 [m(t),\ dW(t;v)]$$

$$m_k^2 = \int_0^1 [m(t), m(t)]\, dt$$

**we obtain (5.15) after a little arithmetic.**

Finally let us remove the restriction in Theorem 5.2 that J (equivalently $(L+L^*)$) be trace-class, (even though this is true for the applications in Chapters 6, 7 and 8). Thus, let us consider the case where the operator $(L+L^*)$ is NOT necessarily trace-class. First of all, let us note a fact that is independent of whether J is trace-class or not. Let

$$Z(t;\omega) = Y(t;\omega) - \int_0^t M(s) \int_0^s L(\sigma)dY(\sigma;\omega))\, ds$$

Then $Z(t;\omega)$ is a Gaussian process continuous in $t$ with probability one, and further it is a Wiener process. In fact it is immediate that:

$$E\left( \int_0^1 [f(t),\ dZ(t;\omega)] \cdot \int_0^1 [g(t),\ dZ(t;\omega)] \right)$$

$$= [R(I-L)^* f,\ (I-L)^* g] = [f, g]$$

Moreover, let K be defined by:

$$(I-L)^{-1} = I-K$$

Then K is Volterra, with kernel K(t, s) which is continuous, and we have:

$$Y(t;\omega) = Z(t;\omega) - \int_0^t \left( \int_0^s K(s;\sigma)\, dZ(\sigma;\omega) \right) ds$$

Next, J not being trace-class, we cannot use (4.2a); hence, let us proceed circumspectly as follows. For each n, let us approximate L as in Appendix I, by defining a subdivision of the interval [0,1] with the aid of subdivision points $\{t_i\}$:

$$t_i < t_{i+1} \; ; \quad t_0 = 0 \; ; \quad t_n = 1$$

and defining the operator $L_n$ by:

$$L_n f = g; \quad g(t) = M(t_i) \int_0^{t_i} L(s)\, f(s)\, ds \qquad t_i < t < t_{i+1}$$

Then, $L_n$ is trace-class, and the Hilbert-Schmidt norm:

$$|| L_n - L ||^2_{H.S.}$$

goes to zero as

$$\Delta_n = \max\,(t_{i+1} - t_i)$$

goes to zero. Next let

$$(I - L_n)^{-1} = I - K_n$$

Let $K_n(t;s)$ denote the kernel corresponding to $K_n$. Define

$$y_n\,(t;\omega) = \int_0^t K_n(t;s)\, dZ(s;\omega)$$

which is a step function for each $\omega$, omitting a set of measure zero. Define next the approximating process:

$$Y_n(t;\omega) = Z(t;\omega) - \int_0^t y_n(s;\omega)\, ds$$

Then $Y_n(t;\omega)$ is continuous in $t$ for almost all $\omega$, and hence

$$\varphi_n(\omega) = Y_n(.;\omega)$$

maps into $\mathscr{C}$. Let $p_n$ denote the corresponding measure induced. Note that

$$E(\int_0^1 f(t),\ dY_n(t;\omega)\ .\ \int_0^1 g(t),\ dY_n(t;\omega) = [(I-K_n)(I-K_n^*) f, g]$$

Let

$$R_n^{-1} = (I-L_n)^* (I-L_n) = I + J_n$$

Then $J_n$ is trace-class, and hence by Theorem 5.2, and corollary the measure is absolutely continuous with respect to Wiener measure, with derivative given by (5.13):

$$H_n(v) = \exp.\ h_n(v)$$

where

$$h_n(v) = (-1/2) \int_0^1 [W^n(t;v),\ W^n(t;v)]\, dt + \int_0^1 [W^n(t;v),\ dW(t;v)] \tag{5.15}$$

where

$$W^n(t;v) = \int_0^t L_n(t;s)\, dW(s;v)$$

where $L_n(t;s)$ is the kernel corresponding to the operator $L_n$. Because strictly speaking the Corollary to Theorem 5.2 assumed the continuity of the kernel and here it is not, we shall actually prove this here. Thus, for each n, let $\phi_{k,n}$ denote the orthonormalized eigenvectors of $J_n$ and $\gamma_{k\,n}$ the corresponding eigen values. Since $J_n$ has a finite dimensional range, of course only a finite number of the eigen values are non-zero. Because $J_n$ is trace-class, we can use (4.2a) for the R-N derivative and obtain:

$$H_n(v) = \exp(-1/2)(\text{Tr.}\ (L_n + L_n^*) + \sum_1^\infty \gamma_{k\,n}\ \zeta_{k\,n}(v)^2)$$

$$= \exp(-1/2) \sum_1^\infty \gamma_{kn}\ \zeta_{kn}(v)^2$$

where

$$\zeta_{k\,n}(v) = \int_0^1 [\phi_{k\,n}(t), dW(t;v)]$$

Proceeding as in the proof of the corollary to Theorem 5.2, we can write:

$$(-1/2)\ \Sigma\ \gamma_{kn}\ \zeta_{k\,n}(v)^2$$

$$= (-1/2) \sum_k \sum_m [L_n \phi_{kn}, L_n \phi_{mn}]\ \zeta_{kn}(v)\ \zeta_{m\,n}(v)$$

$$+ \sum_k \sum_m [L_n \phi_{k\,n}, \phi_{m\,n}]\ \zeta_{k\,n}(v)\ \zeta_{m\,n}(v)$$

Since $L_n$ itself is trace class, and continuity of the kernel is not necessary for this, we obtain from (5.9) that the second series

$$= \int_0^1 [W^n(t;v), dW(t;v)]$$

while the first series readily yields

$$(-1/2) \int_0^1 [W^n(t;v), W^n(t;v)]$$

thus completing the required proof. Next let us note that as n goes to infinity, the second term in (5.15) converges in the mean square sense to:

$$\int_0^1 [W_1(t;v), dW_1(t;v)]$$

because this is the way construct the above integral. Next

$$E\left(\left|\int_0^1 [W^n(t;v), W^n(t;v)]dt - \int_0^1 [W_1(t;v), W_1(t;v)]dt\right|\right)$$

$$= E\left(\left|\int_0^1 [W^n(t;v) - W_1(t;v), W^n(t;v) + W_1(t;v)]\, dt\right|\right)$$

$$\leqslant \left(\int_0^1 E\left(\|W^n(t;v) - W_1(t;v)\|^2\right) dt \int_0^1 E\left(\|W^n(t;v) + W_1(t;v)\|^2\right) dt\right)^{1/2}$$

and goes to zero with $n$, since

$$E(\|W^n(t;v) - W_1(t;v)\|^2) = \int_0^t \|L_n(t;s) - L(t;s)\|^2 ds$$

$$E(\|W^n(t;v) + W_1(t;v)\|^2) = \int_0^t \|L_n(t;s) + L(t;s)\|^2 ds$$

Hence a subsequence of (5.15) converges with probability one to:

$$h(v) = (-1/2)\int_0^1 [W_1(t;v), W_1(t;v)]\, dt + \int_0^1 [W_1(t;v), dW(t;v)] \tag{5.16}$$

We next wish to show that

$$H(v) = \exp h(v)$$

is the R-N derivative sought. For this we shall show that $H_n(v)$ are uniformly integrable by obtaining the estimate that

$$\sup_n E(H_n(v)^{1+\alpha}) < \infty \tag{5.17}$$

for some $\alpha$, $0 < \alpha < 1$. First of all, from (4.2), we know that for each $n$, and $\alpha$, such that

$$(1 + (1+\alpha)\gamma_n) > 0 \qquad \gamma_n = \min_k \gamma_{kn} \tag{5.18}$$

we have

$$E(H_n(v)^{1+\alpha}) \leq \exp k(\alpha) \, ||J_n||^2_{H.S.}$$

(where $k(\alpha)$ also depends on $\gamma_n$).

Since $||J_n||_{H.S.}$ is bounded, it is enough to find $\alpha$ such that (5.18) holds for all $n$ sufficiently large. Now $J_n$ converges in the Hilbert-Schmidt norm to J; and for any $\phi$ of norm one, we have:

$$[J_n\phi, \phi] = [(J_n - J)\phi, \phi] + [J\phi, \phi]$$

$$\geq [(J_n - J)\phi, \phi] + \text{smallest eigen value of } J$$

and since the first term is

$$\leq ||J_n - J||_{H.S.}$$

independent of $\phi$, and can be made as small as we wish for large enough $n$. Hence for all $n$ sufficiently large,

$$\gamma_n \geq \epsilon + \text{smallest eigen value of } J > -1 + |\delta|, \quad |\delta| > 0$$

since $|\epsilon|$ can be made smaller than any desired number. This shows that we can indeed find a number $\alpha$, $0 < \alpha < 1$, such that (5.18) holds for all $n$, sufficiently large, and hence we obtain (5.17). Since (subsequence) $h_n(v)$ converges to $h(v)$, so does $H_n(v)$ to $H(v)$. Hence uniform integrability now yields that (working from now on with the convergent subsequence):

$$\int_B | H_n(v) - H(v) | \, dp_W$$

goes to zero for any Borel set B.

Next for any set A of the form:

$$A = [v | \int_0^1 [f(t), dW(t;v)] \in U]$$

where U is a linear Borel set,

$$p_{\varphi}(A) = p_W(\varphi^{-1}(A))$$

$$= \int_U G(x;\sigma)\, dx; \ \sigma^2 = [Rf, f] = \|(I-L)^{-1} f\|^2$$

where $G(x;\sigma)$ denotes the Gaussian density with zero mean and variance $\sigma^2$.

Let

$$\sigma_n^2 = \|(I-L_n)^{-1} f\|^2$$

Then

$\sigma_n^2$ clearly converges to $\sigma^2$

and is bounded away from zero. Hence

$$p_{\varphi}(A) = \int_U G(x;\sigma) dx$$

$$= \text{Lim} \int_U G(x;\sigma_n) dx$$

$$= \text{Lim } p_n(A)$$

$$= \int_A e^{h(v)}\, dp_W$$

As we know this is enough to prove the statement for every Borel set. This proves the extension sought. The non-zero mean case can be treated similarly.

Example: Note in particular that for the general case of (4.3a) the Radon-Nikodym derivative is thus given by:

$$H(\omega) = \exp \left\{ (-1/2) \int_0^1 [A(t)\, W(t;\omega)\, ,\ A(t)\, W(t;\omega)] + \int_0^1 [A(t) W(t;\omega), dW(t;\omega)] \right\} \ ..$$

Example  Suppose in (2.14), we take the special case where:

(i) $D(s)D(s)^* = $ Identity matrix.

(ii) $B(s)D(s)^* = 0, \quad 0 \leq s \leq 1$

Then $R$ in (2.19) has the form

$$R = I + K^*K$$

This implies that

$$J = R^{-1} - I = -R^{-1}(K^*K)$$

and $J$ is trace-class, since $K^*K$ is. By a theorem of Krein this implies that

$$I + J = (I-L)^*(I-L)$$

where $L$ is Volterra. See Appendix II for a proof. After we do Kalman filtering, we shall see an alternate method for actually obtaining $L$. Once we can represent $L$ as:

$$Lf = g; \; g(t) = M(t)\int_0^t L(s)f(s)ds$$

we can apply Theorem 5.2 directly to obtain the R-N derivative in this case.

More generally, if R has the form

$$R = I + T$$

where $T$ is given by:

$$Tf = g \quad g(t) = \int_0^1 T(t;s)f(s)ds$$

and $T(t;s)$ is continuous on $0 \leq s, t \leq 1$ and $T$ is non-negative definite, then $T$ is trace-class and hence so is $J$. The Krein theorem then yields the operator $L$ needed in the Corollary, and hence the Radon-Nikodym derivative is given by (5.15).

# CHAPTER VI

## LINEAR RECURSIVE ESTIMATION

A fundamental problem in communication and control is the filtering (estimation) problem which can be stated as follows: Given an observed process

$$y(t;\omega)$$

which has the structure:

$$y(t;\omega)=s(t;\omega) + n(t;\omega)$$

where $s(t;\omega)$ is the "signal process" and $n(t;\omega)$ the "observation error" or "noise" modelled as a stochastic process, find the "best estimate" of the signal $s(t;\omega)$ from the available data at time $t$:

$$y(\sigma;\omega), \quad a \leq \sigma \leq t$$

We shall be concerned with the case where the best estimate is taken as the best "mean square" estimate (see Chapter 3) - in other words, we wish to calculate the conditional expectation:

$$\hat{s}(t;\omega) = E[s\,(t;\omega)\,|\,y(\sigma;\omega),\; a \leq \sigma \leq t]$$

Furthermore we shall be concerned only with the case where the processes are (jointly) Gaussian, so that in particular the conditional estimate is a linear transformation of the observation. In fact, confining ourselves to zero mean processes, we know that (see [2], [40] for example)

$$\hat{s}(t;\omega) = \int_a^t k(t;\sigma)\, y(\sigma;\omega)d\sigma$$

where the weight function $k(t;\sigma)$ is such that

$$E((s(t;\omega)-\hat{s}(t;\omega))(y(\sigma;\omega))^*) = 0, \quad a \leq \sigma \leq t \ .$$

The main problem then is that of finding $k(t;\sigma)$. When the processes are stationary and their spectra are known and $a = -\infty$, this problem was solved by Kolmogorov - Wiener - see Doob [2]. However such a problem rarely occurs in practice. Here we shall be concerned with the case where $s(t;\omega)$ can be characterized by a stochastic differential system and $n(t;\omega)$ is a "white noise" (or "derivative of Wiener process" loosley speaking). In that case it turns out that the estimate $\hat{s}(t;\omega)$ can also be characterized by a differential system - and hence we have: "recursive estimation". Not only is the theory richer by virtue of the differential structure imposed; it also turns out to be the kind of structure essential in practical application (e.g. "trajectory estimation" - see [21, 22]) as well as stochastic control theory (see Chapter 7). It is essential in the theory that the noise process, in Engineering terms, is "white Gaussian". For a precise formulation we have to use Wiener processes, and the "integrated" version. Thus the problem takes the following form: The observation process has the form:

$$Y(t;\omega) = \int_0^t C(s)\, x(s;\omega)ds + \int_0^t D(s)dW(s;\omega)$$

where $W(t;\omega)$ is a Wiener process, and the second term corresponding to the noise is such that

$$D(s)\, D(s)^* > 0.$$

The first term corresponds to the signal, and $x(t;\omega)$ satisfies the stochastic equation:

$$x(t;\omega) = \int_0^t A(s)\, x(s;\omega) ds + \int_0^t B(s)\, dW(s;\omega).$$

The problem is to "find" (or rather characterize)

$$\hat{x}(t;\omega) = E[x(t;\omega) \mid Y(s;\omega), \quad 0 \leq s \leq t]$$

Note that the best estimate of the signal

$$C(t)\, x(t;\omega)$$

is of course

$$C(t)\, \hat{x}(t;\omega).$$

The problem was first formulated in this form by Bucy-Kalman and solved in largely formal manner and is since associated with their names. The approach here is different from the original versions based on Wiener-Hopf equation analysis, as well as the more recent version due to Wonham [7]. The main idea is to exploit the simpler problem of estimating one Martingale from another. It would be useful to begin with an outline.

The salient steps in our approach are:

$$\text{i)}\ Y(t) - \int_0^t C(s)\hat{x}(s;\omega)ds = Z_0(t;\omega)$$

is a Martingale (this result is known, see [7], [16]).

But it is also true that

$$\text{ii)}\ \hat{x}(t;\omega) - \int_0^t A(s)\hat{x}(s;\omega)ds = Z_s(t;\omega)$$

is also a Martingale.

iii) Given the two Martingales, $Z_0(t;\omega)$, $Z_s(t;\omega)$, the best mean square estimate $\hat{Z}_s(t;\omega)$ of $Z_s(t;\omega)$ from $Z_0(\sigma;\omega)$, $\sigma \leq t$ is given by:

$$\hat{Z}_s(t;\omega) = \int_0^t \lambda_{12}(s)\, dZ_0(s;\omega)$$

iv) we give a simple proof of the equivalence of the sigma-algebras generated by $Z_0(t;\omega)$ and $Y(t;\omega)$ so that $Z_s(t;\omega) = \hat{Z}_s(t;\omega)$, and substituting this into (ii) yields the emation sought

We begin with a fundamental property of Martingales. (Cf. Nelson, Doob).

<u>Lemma</u> Let $Z_i(t;\omega)$, $i = 1, 2$ denote two Martingales with respect to the growing sigma algebra $\mathscr{F}(t)$, and let:

$$E(\| Z_i(1;\omega) - Z_i(0;\omega)\|^2) < \infty \ , \quad i = 1, 2 \ \ldots . \qquad (6.$$

Suppose that for $i,j$ fixed, : $0 \leq t < 1$, :

$$\lim_{\Delta \to 0} \frac{1}{\Delta} E((Z_i(t+\Delta;\omega) - Z_i(t;\omega))(Z_j(t+\Delta;\omega) - Z_j(t;\omega))^* | \mathscr{F}(t)) = P_{ij}(t) \qquad (6.2)$$

where the convergence of the random variable on the left is in the mean order one $(L_1)$, and it is assumed that $P_{ij}(t)$ is continuous in $t$, $0 \leq t \leq 1$. Then for $0 \leq s < t \leq 1$,

$$E\left(\left(\int_s^t dZ_i(\sigma;\omega)\right)\left(\int_s^t dZ_j(\sigma;\omega)\right)^* \Big| \mathscr{F}(s)\right) = \int_s^t P_{ij}(\sigma)d\sigma \qquad (6.3)$$

Proof Let $0 \leq a < 1$, and define

$$\Lambda_{ij}(t) = \left(\int_a^t dZ_i(s;\omega)\right)\left(\int_a^t dZ_j(s;\omega)\right)^*, \quad a \leq t < 1$$

Observe that for any $\Delta > 0$, $t + \Delta < 1$, we have:

$$E(\Lambda_{ij}(t+\Delta) - \Lambda_{ij}(t)) | \mathscr{F}(a)) = E[E[(\int_t^{t+\Delta} dZ_i(s;\omega))(\int_t^{t+\Delta} dZ_j(s;\omega))^* | \mathscr{F}(t)] \mathscr{F}(a)]$$

where we have used the fact that:

$$E[[(Z_i(t+\Delta;\omega) - Z_i(t;\omega))(Z_j(t;\omega) - Z_j(a;\omega))^*] | \mathscr{F}(t)] = 0$$

We can now follow the argument of Nelson ( 8 ) ; Let $\epsilon > 0$ be given. Let $J$ denote the set of points in $[a, 1]$ such that for $t$ in $J$:

$$E(|E(\Lambda_{ij}(t)|\mathscr{F}(a)) - \int_a^t P_{ij}(s)ds|) \leq \epsilon\ (t-a) \qquad (6.4)$$

Clearly $J$ contains the point $a$. Also $J$ is closed. Now we shall show that if $t$ is any point for which ( 6.4 ) holds, then it will hold for $t+\delta$, $0<\delta<\Delta$, for some $\Delta>0$. For this, we note that

$$E(\Lambda_{ij}(t+\delta)\,|\,\mathscr{F}(a)) = E(E[\Lambda_{ij}(t+\delta)-\Lambda_{ij}(t)\,|\,\mathscr{F}(t)]\,|\,\mathscr{F}(a)) + E(\Lambda_{ij}(t)\,|\,\mathscr{F}(a))$$

and let us choose $\Delta_1$ such that for all $\delta<\Delta_1$,

$$E(|(E(\Lambda_{ij}(t+\delta)-\Lambda_{ij}(t)/\mathscr{F}(t)) - \delta P_{ij}(t))|) < (\epsilon/2)\,\delta$$

Next let us choose $\Delta_2$ such that for all $\delta<\Delta_2$,

$$\left|\int_t^{t+\delta} P_{ij}(s)ds - \delta P_{ij}(t)\right| < (\epsilon/2)\delta$$

Choosing $\Delta$ to be the minimum of $\Delta_1$ and $\Delta_2$, we have: for $0\le\delta\le\Delta$

$$E\left|(E(\Lambda_{ij}(t+\delta)-\Lambda_{ij}(t)/\mathscr{F}(a)) - \int_t^{t+\delta} P_{ij}(s)ds)\right|$$

$$= \quad E\left|(E(\Lambda_{ij}(t+\delta)-\Lambda_{ij}(t)/\mathscr{F}(t) - \int_t^{t+\delta} P_{ij}(s)ds\,/\,\mathscr{F}(a)\,)\right|$$

$$\le \quad (\epsilon)\,\delta.$$

Hence (6.4) follows for $t+\delta$, $0<\delta<\Delta$. In particular this is true for $t=a$. Suppose now that the upperbound of $t$ such that (6.4) holds for $[a,t]$ is $t_o$, say. Then $t_o$ must belong to $J$ since $J$ is closed. But if $t_o$ is not equal to $1$, we will have a contradiction because it can be extended by a non-zero amount. Since $\epsilon$ is arbitrary, it is clear that:

$$E[\Lambda_{ij}(t)\,|\,\mathscr{F}(a)] = \int_a^t P_{ij}(s)ds$$

which is clearly enough to prove the Lemma.

Corollary: Assume now that the Martingales $Z_i(s;\omega)$ are Gaussian, $0 \le s \le 1$, and that

$$Z_i(0;\omega) = 0, \; i = 1,2$$

Assume further that (6.1) holds, and that (6.2) holds for $i = j = 2$, and for $i = 1, j = 2$. Let $\mathscr{B}_2(t)$ the smallest sigma algebra generated by $Z_2(s;\omega)$, $s \le t$. Then

$$E(Z_1(t;\omega) \mid \mathscr{B}_2(t)) = \int_0^t r_{12}(s)dZ_2(s;\omega) \qquad 0 < t \le 1 \tag{6.5}$$

where $r_{12}(s)$ is defined as the limit:

$$r_{12}(s) = \lim_{\epsilon \to 0} P_{12}(s)\,(P_{22}(s) + \epsilon I)^{-1} \quad \text{a.e.} \quad 0 < s < 1 \tag{6.6}$$

Proof It should be emphasized that it is not assumed that (6.2) holds for $i = j = 1$. First let us prove that for any Lebesgue measurable matrix function $f(.)$, having the same dimension as $P_{12}(s)$, such that

$$\int_0^1 \text{Tr. } f(s)^* f(s) P_{22}(s)ds < \infty \tag{6.7}$$

we have that:

$$E((\int_0^1 dZ_1(s;\omega))\,(\int_0^1 f(s)dZ_2(s;\omega))^*) = \int_0^1 P_{12}(s)f(s)^*ds \tag{6.8}$$

For this it is enough to note that for $0 < s < t < 1$,

$$E(\int_0^1 dZ_1(s;\omega))\,(Z_2(t;\omega) - Z_2(s;\omega))^*)$$

$$= \quad E(Z_1(t;\omega) - Z_1(s;\omega))(Z_2(t;\omega) - Z_2(s;\omega))^*) = \int_s^t P_{12}(\sigma)d\sigma$$

from (6.3) ; hence it is immediate that (6.8) will hold for simple functions, and hence by the usual limiting arguments to any $f(.)$ satisfying (6.7) . Next if

$$L(\epsilon;t) = P_{12}(t)(P_{22}(t) + \epsilon I)^{-1}$$

we have that: $L(\epsilon;t)$ satisfies (6.7), and that:

$$0 \leq E(\| Z_1(1;\omega) - \int_0^1 L(\epsilon;t)\, dZ_2(t;\omega) \|^2)$$

$$= E\| Z_1(1;\omega)\|^2 - \text{Tr.} \int_0^1 P_{12}(t)(P_{22}(t) + \epsilon I)^{-1}(P_{22}(t) + 2\epsilon I)$$

$$(P_{22}(t) + \epsilon I)^{-1} P_{12}(t)^* dt$$

By Fatou's Lemma, it follows that: $r_{12}$ satisfies (6.7). Hence the integral:

$$\int_0^t r_{12}(s) dZ_2(s;\omega) \qquad 0 < t < 1$$

is well defined. Let $f(.)$ satisfy (6.7). Then

$$E(((Z_1(t;\omega) - \int_0^t r_{12}(s)dZ_2(s;\omega)) (\int_0^t f(s)dZ_2(s;\omega))^*)$$

$$= \int_0^t (P_{12}(s)f(s)^* - r_{12}(s)P_{22}(s)f(s)^*)ds$$

But

$$r_{12}(s)P_{22}(s) = (P_{12}(s) \lim_{\epsilon \to 0} (P_{22}(s) + \epsilon I)^{-1} P_{22}(s)$$

$$= P_{12}(s)$$

which is trivially true at a point where

$$P_{22}(s) > 0$$

and otherwise because if

$$P_{22}(s)\, x = 0$$

so is

$$P_{12}(s)x = 0$$

as can be verified from (6.2). Hence

$$Z_1(t;\omega) - \int_0^t r_{12}(s)dZ_2(s;\omega)$$

is uncorrelated with, and being Gaussian, independent of,

$$\int_0^t f(s)dZ_2(s;\omega)$$

for every $f(\cdot)$ satisfying (6.7). But since the latter generate $\mathscr{B}_2(t)$, (6.5) follows.

Let us consider now the linear stochastic equation (cf.(2.16)):

$$x(t;\omega) = \int_0^t A(s)\, x(s;\omega)\, ds + \int_0^t B(s)dW(s;\omega) \tag{6.9}$$

$$Y(t;\omega) = \int_0^t C(s)x(s;\omega)ds + \int_0^t D(s)dW(s;\omega) \tag{6.10}$$

where we shall for simplicity assume that all coefficients are continuous. Further we shall assume:

$$D(s)D(s)^* > 0 \quad \text{on} \quad [0,1]$$

Note that this implies that

$$b(s) = \sqrt{(D(s)D(s)^*)^{-1}}$$

is continuous. Then as we have seen, defining:

$$\hat{Y}(t;\omega) = \int_0^t b(s)dY(s;\omega)$$

we have:

$$\hat{Y}(t;\omega) = \int_0^t b(s)C(s)x(s;\omega)ds + \hat{W}(t;\omega)$$

where $\hat{W}(s;\omega)$ is now a Wiener process. Let $\hat{W}(t;\omega)$ be measurable the growing sigma-algebra $\mathscr{F}(t)$. Then of course $x(t;\omega)$, $Y(t;\omega)$, $\hat{Y}(t;\omega)$, $\hat{W}(t;\omega)$ are all measurable $\mathscr{F}(t)$. Let

$$\mathscr{B}_Y(t) = \text{sigma algebra generated by } Y(s;\omega),\ s \leq t$$

and similarly define $\mathscr{B}_{\hat{Y}}(t)$. Then since

$$Y(t;\omega) = \int_0^t (b(s))^{-1} d\hat{Y}(s;\omega)$$

we have that

$$\mathscr{B}_Y(t) = \mathscr{B}_{\hat{Y}}(t)$$

Since we have that

$$E(|x(t;\omega)|^2) < \infty$$

we can define the conditional expectation:

$$\hat{x}(t;\omega) = E(x(t;\omega) \,|\mathscr{B}(t))$$

where from now on we write

$$\mathscr{B}(t) \text{ for } \mathscr{B}_Y(t) = \mathscr{B}_{\hat{Y}}(t)$$

Observe now that:

$$E(x(t;\omega)|\mathscr{B}(s)) = E(E(\phi(t)\,\phi(s)^{-1}x(s;\omega)+\phi(t)\int_s^t \phi(\sigma)^{-1}B(\sigma)dW(\sigma;\omega))\,|\,\mathscr{F}(s))|\mathscr{B}(s))$$

$$= \phi(t)\,\phi(s)^{-1}\hat{x}(s;\omega)$$

But the leftside is a Martingale in $s$, $s < t$, for fixed $t$, and converges with probability one as $s$ goes to $t$. But $Y(t;\omega)$ being continuous with probability one, we note that

$\mathscr{B}(t)$ = smallest sigma-algebra containing $\mathscr{B}(s)$ for every $s < t$

and hence the limit must be $\hat{x}(t;\omega)$. Thus $\hat{x}(t;\omega)$ is continuous from below with probability one. Hence by a theorem of Doob ([2 ], Theorem 2.6, p. 61, and remark on p. 62), there is a process equivalent to $\hat{x}(t;\omega)$ which is jointly measurable in $t$ and $\omega$, and we may clearly redefine $\hat{x}(t;\omega)$ to be this process. And since

$$E(|\hat{x}(t;\omega)|^2) \leq E(|x(t;\omega)|^2) < \infty$$

it follows that the integral

$$\int_0^t A(s)\,\hat{x}(s;\omega)ds\ , \quad t \leq 1$$

is well-defined (converges a.s.).

Lemma: The process:

$$\hat{x}(t;\omega) - \int_0^t A(s)\,\hat{x}(s;\omega)\,ds = Z_s(t;\omega) \quad 0 \leq t \leq 1$$

is a Gaussian Martingale. Moreover

$$E(|Z_s(t;\omega)|^2) < \infty$$

Proof Now for $t > s$, we have:

$$E(\hat{x}(t;\omega)|\mathscr{B}(s)) = E(x(t;\omega)\mid\mathscr{B}(t))|\mathscr{B}(s))$$

$$= E(x(t;\omega)|\mathscr{B}(s)) \ldots\ldots \tag{6.11}$$

Hence

$$E(Z_s(t;\omega) - Z_s(s;\omega)\ |\mathscr{B}(s)) = E((x(t;\omega) - x(s;\omega))|\mathscr{B}(s))$$

$$- E\left(\int_s^t A(\sigma)\,\hat{x}(\sigma;\omega)d\sigma\,|\mathscr{B}(s)\right)$$

$$= E\left(\int_s^t B(\sigma)dW(\sigma;\omega)|\mathscr{B}(s)\right) + E\left(\int_s^t A(\sigma)(x(\sigma;\omega)-\hat{x}(\sigma;\omega))d\sigma\,|\mathscr{B}(s)\right)$$

But the first term is zero because $\mathscr{F}(s) \supset \mathscr{B}(s)$, and $\int_0^t B(\sigma)dW(\sigma,u)$ is a Martingale; the second term is zero, using (6.11).

Lemma: Let

$$Z_0(t;\omega) = Y(t;\omega) - \int_0^t C(s)\hat{x}(s;\omega)ds$$

Then $Z_0(t,u)$ is a Gaussian martingale. Moreover:

$$\lim_{\Delta \to 0} (1/\Delta)E\left(\int_t^{t+\Delta} dZ_0(s;\omega)\left(\int_t^{t+\Delta} dZ_0(s;\omega)^*\right) \Big| \mathscr{B}(t)\right) = D(t)D(t)^* \tag{6.12}$$

Proof We have only to note that:

$$Z_0(t;\omega) = Y(t;\omega) - \int_0^t C(s)x(s;\omega)\,ds + \int_0^t C(s)(x(s;\omega) - \hat{x}(s;\omega))ds$$

$$= \int_0^t D(s)dW(s;\omega) + \int_0^t C(s)(x(s;\omega) - \hat{x}(s;\omega))ds$$

so that

$$Z_0(t;\omega) - Z_0(s;\omega) = \int_s^t D(\sigma)dW(\sigma;\omega) + \int_s^t C(\sigma)(x(\sigma;\omega) - \hat{x}(\sigma;\omega))d\sigma$$

and just as in the previous lemma,

$$E((Z_0(t;\omega) - Z_0(s;\omega)) \,|\, \mathscr{B}(s)) = 0$$

Let us use the notation:

$$e(s;\omega) = x(s;\omega) - \hat{x}(s;\omega) \tag{6.13}$$

and observe that

$$E\left(\left\| \int_t^{t+\Delta} C(s)\, e(s;\omega)ds \right\|^2 \Big| \mathscr{B}(t)\right) = 0\,(\Delta^2)$$

since

$$E(|e(s;\omega)|^2) \le E(|x(s;\omega)|^2)$$

and is bounded in $0 < s < 1$. Again:

$$E(\| \int_t^{t+\Delta} D(s)dW(s;\omega) \|^2 | \mathscr{B}(t)) = 0(\Delta)$$

Hence clearly:

$$\frac{1}{\Delta} E((\int_t^{t+\Delta} dZ_0(s;\omega))(\int_t^{t+\Delta} dZ_0(s;\omega))^* | \mathscr{B}(t)) = \frac{1}{\Delta}\int_t^{t+\Delta} D(s)D(s)^* ds + 0(\Delta^{1/2})$$

Hence (6.12) follows.

Lemma: Let

$$E(e(t;\omega)\ \dot{e}(t;\omega)^*) = P(t)$$

Then

$$\lim_{\Delta \to 0} (1/\Delta) E(\int_t^{t+\Delta} dZ_s(\sigma;\omega))(\int_t^{t+\Delta} dZ_0(s;\omega))^*) | \mathscr{B}(t))$$

$$= P(t)C(t)^* + B(t)D(t)^* \quad \text{a.e.} \qquad (6.14)$$

Proof Since $Z_0(s;\omega)$ satisfies (6.3) and $Z_s(t;\omega)$ satisfies (6.2), it only remains to calculate (6.14). We have seen that:

$$Z_s(t+\Delta;\omega) - Z_s(t;\omega) = \int_t^{t+\Delta} A(s)e(s;\omega)ds + \int_t^{t+\Delta} B(s)dW(s;\omega)$$

$$- e(t+\Delta;\omega) + e(t;\omega)$$

It is immediate that

$$|E((\int_t^{t+\Delta} A(s)\, e(s;\omega)ds\, (\int_t^{t+\Delta} dZ_0(s;\omega))^* | \mathcal{B}(t))| = 0(|\Delta|^{3/2})$$

Next for any $\Delta > 0$, $e(t+\Delta;\omega)$ is uncorrelated with $Y(s;\omega)$, $s \leq (t+\Delta)$, and hence with $Z_0(s;\omega)$, $s \leq (t+\Delta)$. It is also uncorrelated with (and hence also independent of) the random variables generating $\mathcal{B}(t)$. Hence

$$E(e(t+\Delta;\omega)\, (\int_t^{t+\Delta} dZ_0(s;\omega))^* | \mathcal{B}(t)) = E(e(t+\Delta;\omega))E(\int_t^{t+\Delta} dZ_0(s;\omega))^* = 0$$

Since

$$Z_0(s;\omega) = \int_0^s C(\sigma)e(\sigma;\omega)d\sigma + \int_0^s D(\sigma)dW(\sigma;\omega)$$

we have:

$$E(e(t;\omega)\, (\int_t^{t+\Delta} dZ_0(s;\omega))^* | \mathcal{B}(t)) = \int_t^{t+\Delta} E(e(t;\omega)e(s;\omega)^* | \mathcal{B}(t))C(s)^* ds$$

$$= \int_t^{t+\Delta} E(e(t;\omega)\, e(s;\omega)^*)\, C(s)^* ds$$

and

$$E(\int_t^{t+\Delta} B(s)dW(s;\omega)\, (\int_t^{t+\Delta} dZ_0(s;\omega))^* | \mathcal{B}(t)) = \int_t^{t+\Delta} B(s)D(s)^* ds + 0(\Delta^{3/2})$$

since

$$E(\int_t^{t+\Delta} B(s)dW(s;\omega)\, (\int_t^{t+\Delta} C(s)e(s;\omega)ds)^* | \mathcal{B}(t)) = 0(\Delta^{3/2})$$

Hence (6.14) follows by taking the limit as $\Delta$ goes to zero. Note that a.e. in (6.14) is necessary since we cannot (at this stage) assert that $e(t;\omega)$ is continuous in $t$. Eventually we shall show that it is. (Cf. (6.12)).

Lemma: Under the assumption that

$$D(s)D(s)^* > 0, \quad 0 \leq s \leq 1$$

we can write:

$$\hat{x}(t;\omega) = \int_0^t k(t;s)\, dY(s;\omega)\ ; \quad 0 \leq t \leq 1 \tag{6.15}$$

where

$$\int_0^1 \int_0^t |k(t;s)|^2 ds\, dt < \infty \tag{6.16}$$

Proof Let us first note that

$$\int_0^t f(s)dY(s;\omega) = \int_0^t (f(s)D(s) + h(s))\, dW(s;\omega)$$

where

$$h(s) = \int_s^t f(\sigma)C(\sigma)\phi(\sigma)d\sigma \quad \phi(s)^{-1}\ B(s)$$

Define the operator $L$ by

$$Lf = g\ ; \quad g(s) = f(s)D(s) + \int_s^t f(\sigma)C(\sigma)\phi(\sigma)d\sigma\ \phi(s)^{-1}\ B(s)$$

Then it follows that

$$E((\int_0^t f(s)dY(s;\omega))\ (\int_0^t q(s)\, dY(s;\omega))^*) = \int_0^t p(s)\, q(s)^* ds \tag{6.17}$$

where

$$L^*Lf = p$$

Here f(.) is an m-by-q matrix function, and p(.) is m-by-q consistent with the choice of dimensions on p. 30. Specifically:

$$L^*h = p \;\; ; \; p(s) = h(s)D(s)^* + \int_0^s h(\sigma)B(\sigma)^*\phi(\sigma)^{*-1}d\sigma \; \phi(s)^* C(s)^* \tag{6.18}$$

where h(.) is m-by-n and p(.) is m-by-q.

Next

$$E(x(t;\omega) \; (\int_0^t q(s)dY(s;\omega))^*)$$

$$= E((\phi(t)\int_0^t \phi(s)^{-1}B(s)dW(s;\omega)) \; (\int_0^t q(s)dY(s;\omega))^*)$$

$$= \int_0^t r(s) \; q(s)^* \; ds$$

where

$$r = L^* v \text{ and } v(s) = \phi(t) \; \phi(s)^{-1} \; B(s), \quad 0 \le s \le t \tag{6.19}$$

Now

$$E(x(t;\omega) - \int_0^t k(t;s)dY(s;\omega)) \; (\int_0^t q(s)dY(s;\omega))^*) = \int_0^t z(s) \; q(s)^* \; ds$$

where

$$L^*L \; k - L^* v = -z$$

where k stands for the function $k(t;\cdot)$. Hence for (6.15) to hold it is necessary and sufficient that

$$z(s) = 0\,; \quad \text{or,} \quad L^*L\,k = L^*\,v \tag{6.20}$$

But because $D(s)D(s)^*$ is positive, $L^*L$ has a bounded inverse, so that the first part of our result that there exists a function $k(t;s)$ satisfying (6.15) and such that

$$\int_0^t |k(t;s)|^2 ds < \infty$$

for each $0 < t < 1$, is immediate. We need however to show that the double integral in (6.16) makes sense and that it is finite. For this we proceed to a closer examination of (6.20). Thus we note that if $Lf = g$, we can write:

$$g(s) = f(s)D(s) - \int_0^s f(\sigma)\; C(\sigma)\; \phi(\sigma)d\sigma\; \phi(s)^{-1}\; B(s)$$

$$+ \int_0^t f(\sigma)\; C(\sigma)\; \phi(\sigma)d\sigma\;\; \phi(s)^{-1}\; B(s)$$

The point in doing this is that the first two terms are independent of $t$. Again if we denote by $\tilde{L}$ the operator yielding the first two terms:

$$\tilde{L}\,f = g\,; \quad f(s)D(s) - \int_0^s f(\sigma)\; C(\sigma)\; \phi(\sigma)\; d\sigma\; \phi(s)^{-1}\; B(s) = g(s) \tag{6.2.}$$

we have

$$L^*L\,f = L^*\tilde{L}f + \left(\int_0^t f(\sigma)C(\sigma)\;\phi(\sigma)\,d\sigma\right)\left(\phi(s)^{-1}B(s)D(s)^*\right.$$

$$\left. + R(s)\;\phi(s)^*\;C(s)^*\right) \tag{6.2.}$$

where

$$R(s) = \int_0^s \phi(\sigma)^{-1} B(\sigma) B(\sigma)^* \phi(\sigma)^{*-1} d\sigma$$

Also, the function $r(\cdot)$ in (6.19) can be expressed:

$$r(s) = \phi(t) \phi(s)^{-1} B(s)D(s)^* + \phi(t)R(s)\phi(s)^*C(s)^* \tag{6.23}$$

Hence (6.20) can be written:

$$h(t;s) = -\int_0^t k(t;\sigma)C(\sigma)\phi(\sigma) d\sigma \; u(s) + \phi(t) u(s)$$

where

$$h(t;\cdot) = L^*\tilde{L} (k(t;\cdot))$$

$$u(s) = \phi(s)^{-1} B(s) D(s)^* + R(s) \phi(s)^* C(s)^*$$

We now exploit the fact that $h(t;s)$ factors into a function of $t$ and a function of $s$:

$$h(t;s) = (-\int_0^t k(t;\sigma)C(\sigma)\phi(\sigma) d\sigma + \phi(t)) u(s)$$

But since $D(s)D(s)^*$ is assumed positive, we note that $L^*$ has a bounded inverse in $L_2(0, 1)$, and so does $\tilde{L}$; moreover if

$$L^* h = p$$

we have

$$h(s) = h_1(s)D(s)$$

$$p(s)(D(s)D(s)^*)^{-1} = h_1(s) + \int_0^s h_1(\sigma)D(\sigma)\, B(\sigma)^*\phi(\sigma)^{*-1} d\sigma\, \phi(s)^*C(s)^*$$

Since the right side is 'identity plus Volterra operator', we can use a Neumann expansion to find the inverse; and similarly for $\tilde{L}$. But since each of these operations does not involve $t$, it is readily seen that this implies that it is possible to express $k(t;s)$ as

$$k(t;s) = k_1(t)k_2(s) \text{ (where } k_1(\cdot) \text{ is m-by-m)} \tag{6.24}$$

Hence substituting this into (6.20), and taking advantage of the forms in (6.22) and (6.23) we must have:

$$L^*\tilde{L}k_2 = u(\cdot);\ \ u(s) = \phi(s)^{-1}B(s)D(s)^* + R(s)\phi(s)^*C(s)^*,\ 0 < s < 1$$

And $k_1(t)$ must satisfy:

$$\phi(t) - \int_0^t k_1(t)k_2(s)C(s)\phi(s)ds = k_1(t)$$

But since $\phi(t)$ is nonsingular, it follows that both $k_1(t)$ and

$$\left(I + \int_0^t k_2(s)C(s)\phi(s)ds\right)$$

are nonsingular, and hence

$$k_1(t) = \phi(t)\left(I + \int_0^t k_2(s)C(s)\phi(s)ds\right)^{-1} \tag{6.25}$$

from which (6.16) follows. Of course we have obtained more than we sought to prove.

Now we can prove one of the main theorems (7, 11).

Theorem 6.1 Under the assumption that

$$D(s)D(s)^* > 0$$

for every $s$, $0 \leq s \leq 1$, we have for every $t$, $0 \leq t \leq 1$,

$$\mathcal{B}(t) = \text{smallest sigma algebra generated by } \left\{Z_0(s;\omega),\ s \leq t\right\} \tag{6.26}$$

Proof Let us recall that

$$Z_0(s;\omega) = Y(s;\omega) - \int_0^s C(\sigma)\hat{x}(\sigma;\omega)\, d\sigma \tag{6.27}$$

Hence for any q-by-q matrix function $f(\cdot)$ in (appropriate dimensional) $L_2(0,t)$ space, we have:

$$\int_0^t f(s)dZ_0(s;\omega) = \int_0^t f(s)dY(s;\omega) - \int_0^t f(s)C(s)\hat{x}(s;\omega)ds$$

But using (6.15)

$$\int_0^t f(s)C(s)\hat{x}(s;\omega)ds = \int_0^t f(s)C(s)\int_0^s k(s;\sigma)\, dY(\sigma;\omega)ds$$

$$= \int_0^t \left(\int_\sigma^t f(s)C(s)k(s;\sigma)ds\right)dY(\sigma;\omega)$$

Introduce now the operators:

$$H f = g;\ \ g(\sigma) = f(\sigma) - \int_\sigma^t f(s)C(s)k(s;\sigma)ds\ ,\quad 0 \leq \sigma \leq t$$

which maps $L_2(0,t)$ into itself. More importantly, in view of (6.16) it differs from the identity operator by a Volterra operator with a square integrable kernel. Hence H has a bounded inverse.

Hence for any $g(\cdot)$ in $L_2(0,t)$,

$$\int_0^t g(s)dY(s;\omega) = \int_0^t f(s)dZ_0(s;\omega), \quad f = H^{-1} g$$

Hence the random variables

$$\int_0^t g(s)dY(s;\omega)$$

are measurable with respect to the smallest sigma algebra generated by $\{Z_0(s;\omega), s \leq t\}$, and hence $\mathscr{B}(t)$ is contained in that algebra. This proves (6.26) since obviously the rightside of (6.26) is contained in $\mathscr{B}(t)$.

<u>Theorem 6.2</u>

$$Z_s(t;\omega) = \int_0^t (P(s)C^*(s) + B(s)\dot{D}(s)^*)(D(s)D(s)^*)^{-1}dZ_0(s;\omega) \qquad (6.$$

<u>Proof</u> First of all, using (6.5), taking $Z_1(t;\omega)$ therein to be $Z_s(t;\omega)$, and $Z_2(t;\omega)$ to be $Z_0(t;\omega)$, and making use of (6.12), (6.14) we have that the conditional expectation of $Z_s(t;\omega)$ with respect to the smallest sigma algebra generated by $\{Z_0(s;\omega), s \leq t\}$, is given by the right side of (6.28). But this algebra, by Theorem 2.7 is the same as $\mathscr{B}(t)$, and $Z_s(t;\omega)$ is of course measurable with respect to $\mathscr{B}(t)$. Hence (6.28) follows.

Finally we note that:

Corollary 1. Let $\hat{R}(t) = E(\hat{x}(t;\omega)\hat{x}(t;\omega)^*)$

Then $\hat{R}(t)$ is absolutely continuous and,

$$\dot{\hat{R}}(t) = A(t)\hat{R}(t) + \hat{R}(t)A^*(t) + (P(t)C(t)^* + B(t)D(t)^*)(D(t)D(t)^*)^{-1}(C(t)P(t)+D(t)B(t)^*) \tag{6.29}$$

Proof We have only to note that by the theorem,

$$\hat{x}(t;\omega) = \int_0^t A(s)\hat{x}(s;\omega)ds + \int_0^t (P(s)C(s)^* + B(s)D(s)^*)(D(s)D(s)^*)^{-1}dZ_0(s;\omega) \tag{6.30}$$

and the result follows by specializing (2.13), and using (6.12).

Corollary 2. $P(t)$ is absolutely continuous and $P(0) = 0$, and

$$\dot{P}(t) = A(t)P(t)+P(t)A(t)^*+B(t)B(t)^* - (P(t)C(t)^*+B(t)D(t)^*)(D(t)D(t)^*)^{-1}$$
$$(C(t)P(t)+D(t)B(t)^*) \tag{6.31}$$

Proof We have only to note that

$$P(t) = E(x(t;\omega)x(t;\omega)^*) - E(\hat{x}(t;\omega)\hat{x}(t;\omega)^*)$$

and using (2.13) and (6.29), the result follows.

The equations (6.27), (6.30), (6.31) are the Kalman filtering equations.

Problem Let $\Psi(t)$ be a fundamental matrix solution of

$$\dot{\Psi}(t) = (A(t) - K(t)\,C(t))\,\Psi(t)$$

where

$$K(t) = (P(t)C(t)^* + B(t)D(t)^*)\,(D(t)D(t)^*)^{-1}$$

Then the function $k(t;s)$ in (6.15) is given by

$$k(t;s) = \Psi(t)\,\Psi(s)^{-1}\,K(s)$$

Hint: Substitute (6.27) into (6.30).

Note in particular that if $B(t)\,D(t)^* \equiv 0$ (signal and noise independent), then

$$C(t)\,k(t,t)\,D(t)\,D(t)^* = C(t)\,P(t)\,C(t)^* \qquad (6$$

Problem With $\Psi(t)$ as above, we have

$$P(t) = \int_0^t \Psi(t)\,\Psi(s)^{-1}[B(s)B(s)^* - B(s)D(s)^*\,(D(s)D(s)^*)^{-1}D(s)B(s)^* + P(s)\,C(s)^*\,(D(s)D(s)^*)^{-1}C(s)P(s)]\,\Psi(s)^{*-1}\,\Psi(t)^*ds$$

Hint Use (6.31)

Corollary 3 Consider the case of non-zero mean.

$$x(t;\omega) = \int_0^t A(s)x(s;\omega)ds + \int_0^t u(s)ds + \int_0^t B(s)dW(s;\omega) \qquad (6.3$$

$$Y(t;\omega) = \int_0^t C(s)x(s;\omega)ds + \int_0^t v(s)ds + \int_0^t D(s)dW(s;\omega) \qquad (6.3$$

where the functions $u(t)$, $v(t)$ are deterministic and such that

$$\int_0^1 \|u(t)\|^2 dt + \int_0^1 \|v(t)\|^2 dt < \infty$$

and the other quantities are as in equations (6.9) and (6.10).

Let

$$\hat{x}(t;\omega) = E(x(t;\omega)/\mathscr{B}_{\dot{Y}}(s),\ s \leq t)$$

Then we have:

$$\hat{x}(t;\omega) = \int_0^t A(s)\hat{x}(s;\omega)\,ds + \int_0^t u(s)ds + Z_s(t;\omega)$$

$$Z_s(t;\omega) = \int_0^t (P(s)C(s)^* + B(s)D(s)^*)(D(s)D(s)^*)^{-1}\, dZ_0(s;\omega)$$

$$Y(t;\omega) = \int_0^t C(s)\,\hat{x}(s;\omega)\,ds + \int_0^t v(s)ds + Z_0(t;\omega)$$

where $Z_0(t;\omega)$ is a Gaussian martingale with covariance $D(t)D(t)^*$, and $P(\cdot)$ is again determined by equafion (6.31).

<u>Proof</u>

Let

$$z(t;\omega) = \int_0^t A(s)\ z(s;\omega)\,ds + \int_0^t B(s)dW(s;\omega)$$

$$\tilde{Y}(t;\omega) = \int_0^t C(s)z(s;\omega)ds + \int_0^t D(s)dW(s;\omega)$$

Then we know if

$$\hat{z}(t;\omega) = E(z(t;\omega)\,\big|\,\mathscr{B}_{\tilde{Y}}(s),\ s \leq t)$$

we have:

$$\hat{z}(t;\omega) = \int_0^t A(s)\hat{z}(s;\omega)ds + Z_s(t;\omega)$$

$$\tilde{Y}(t;\omega) = \int_0^t C(s)\hat{z}(s;\omega)ds + Z_o(t;\omega)$$

with $Z_s(t;\omega)$, $Z_0(t;\omega)$ as in the statement of the Corollary.

Next let $x_u(t)$ be the unique solution of

$$x_u(t) = A(t)x_u(t) + u(t) \; ; \; x_u(0) = 0$$

Then

$$E(x(t;\omega)) = x_u(t)$$

and

$$z(t;\omega) = x(t;\omega) - E(x(t;\omega))$$

Since

$$Y(t;\omega) = \tilde{Y}(t;\omega) + \int_0^t C(s)x_u(s)ds + \int_0^t v(s)ds$$

we have of course that

$$\mathscr{B}_Y(s) = \mathscr{B}_{\tilde{Y}}(s)$$

Finally

$$\hat{x}(t;\omega) = E((x(t;\omega) - E(x(t;\omega))/\mathscr{B}_Y(s), \; s \leq t) + E(x(t;\omega))$$

$$= E(z(t;\omega)/\mathscr{B}_Y(s), \; s \leq t) + x_u(t)$$

$$= E(z(t;\omega)/\mathscr{B}_{\tilde{Y}}(s), \; s \leq t) + x_u(t)$$

$$= \hat{z}(t;\omega) + x_u(t)$$

Hence substituting into the equation for $\hat{z}(t;\omega)$, we have

$$\hat{x}(t;\omega) = \int_0^t A(s)\hat{x}(s;\omega)ds + \int_0^t u(s)ds + Z_s(t;\omega)$$

$$Y(t;\omega) = \int_0^t C(s)\hat{x}(s;\omega)ds + \int_0^t v(s)ds + Z_0(t;\omega)$$

as required.

Finally let us connect this with the Corollary to Theorem 5.2.

Corollary 4 In (6.34), assume that $D(s)D(s)^* = I$ and (for simplicity) $B(s)D(s)^* = 0$. Then the measure induced by the process $Y(t;\omega)$ defined by equation (6.34) is absolutely continuous with respect to Wiener measure, and the Radon-Nikodym derivative is given by:

$$H(c) = \exp - 1/2\left\{\int_0^1 [\hat{m}(t)-m(t)-h(t), \hat{m}(t)-m(t)-h(t)]dt + 2\int_0^1 [\hat{m}(t)-m(t)-h(t), dW(t;c)]\right\} \tag{6.35}$$

where

$$m(t) = C(t)x_u(t) + v(t)$$

$$\hat{m}(t) = \int_0^t M(t)L(s)m(s)ds \; ; \quad h(t) = \int_0^t M(t)L(s)dW(s;c)$$

$$M(t)L(s) = C(t)\,\phi(t)\,\phi(s)^{-1}\,P(s)C(s)^*$$

$$\dot{\phi}(t) = (A(t)-P(t)C(t)^*C(t))\phi(t)$$

Proof We need only to verify that the operator $J$ in Theorem 5.2 is trace-class. In the notation of Corollary 3, we have that

$$Y(t;\omega)-E[Y(t;\omega)] = \tilde{Y}(t;\omega) = \int_0^t C(s)z(s;\omega)ds + \int_0^t D(s)dW(s;\omega)$$

It is evident from this that the operator R in Theorem 5.2 has the form

$$Rf = g \; ; \; g(t) = \int_0^1 R(t;s)f(s)ds + f(t)$$

where $R(t;s)$ is continuous in $0 \leq s, t \leq 1$ and hence the operator $(R-I)$, being non-nega definite, is trace-class and hence so is

$$J = I - R^{-1} = (R-I)R^{-1}$$

Next

$$\hat{z}(t;\omega) = \int_0^t \phi(t)\phi(s)^{-1} P(s)C(s)^* d\tilde{Y}(s;\omega)$$

where

$$\phi(t) = (A(t) - P(t)C^*(t)C(t))\,\phi(t)$$

From

$$Z_0(t;\omega) = \tilde{Y}(t;\omega) - \int_0^t C(s)\hat{z}(s;\omega)ds$$

we see that

$$\int_0^1 [f(t), dZ_0(t;\omega)] = \int_0^1 [g(t), d\tilde{Y}(t;\omega)]$$

where

$$g = f - L^*f$$

where the operator L is defined by:

$$Lw = j; \; j(t) = \int_0^t C(t)\phi(t)\phi(s)^{-1}P(s)C(s)^*w(s)ds$$

The rest of the statements of the Corollary now follows upon specializing the Corollary to Theorem 5.2. We note incidentally that

$$(1/2)\mathrm{Tr}(L + L^*) = \int_0^1 \mathrm{Tr.}\, C(t)P(t)C(t)^* dt$$

Remark: Let us note that $H(\cdot)$ defines a measurable function on the appropriate dimensional space of continuous functions $C$ with Wiener measure on Borel sets. Now $Y(t;\omega)$ maps $\Omega$ into $C$ and is of course measurable. Hence $H(Y(\cdot;\omega))$ is a measurable function of $\omega$. It is called the 'likelihood functional'. We can express $H(Y(\cdot;\omega))$ in a simpler and more suggestive way; in the notation of Corollary 3 we can express $h(t)$ as:

$$= C(t)\hat{x}(t;\omega) - C(t)\, x_u(t) + \hat{m}(t)$$

so that†

$$H(Y(\cdot;\omega)) = \exp - 1/2\left\{\int_0^1 [v(t) + C(t)\hat{x}(t;\omega),\ v(t) + C(t)\hat{x}(t;\omega)] - 2\int_0^1 [v(t) + C(t)\hat{x}(t;\omega),\ dY(t;\omega)]\right\} \qquad (6.36)$$

where

$$\hat{x}(t;\omega) = \int_0^t A(s)\,\hat{x}(s;\omega)\,ds + \int_0^t u(s)\,ds + \int_0^t P(s)C(s)^* (D(s)D(s)^*)^{-1} (dY(s;\omega) - C(s)\hat{x}(s;\omega)d - v(s)ds)$$

---

† For a heuristic interpretation, see 'Supplementary Notes'.

Time-Invariant Systems: Asymptotic Behavior:

Let us now specialize to the case where the system is 'time-invariant': that is, where the matrices $A(t)$, $B(t)$, $C(t)$, $D(t)$ are all constant, independent of $t$, and let us denote them by $A, B, C, D$. Of particular interest then is what happens to $P(t)$, $R(t)$ as $t$ goes to infinity - the situation in regard to

$$R(t) = E[x(t;\omega)\, x(t;\omega)^*]$$

is straight-forward; and in fact, motivates the question regarding the others. First of all, note that we have (equation (2.13)):

$$\dot{R}(t) = A R(t) + R(t) A^* + BB^*; \quad R(0) = 0 \tag{2.13}$$

Also, this can be 'solved' explicitly, (being a linear equation), as:

$$R(t) = \int_0^t e^{A(t-s)}\, BB^*\, e^{A^*(t-s)}\, ds = \int_0^t e^{As}\, BB^*\, e^{A^*s} ds \tag{6.37}$$

Note that from this we also have:

$$\dot{R}(t) = e^{At}\, BB^*\, e^{A^*t} \geq 0$$

Now (6.37) converges as $t \longrightarrow \infty$ if the eigen-values of $(A + A^*)$ are all negative, strictly less than zero. For

$$\frac{d}{dt}[e^{At}x, e^{At}x] = [(A + A^*)\, e^{At}x, e^{At}x] \leq \lambda[e^{At}x, e^{At}x]$$

where $\lambda$ is the largest eigen-value of $(A+A^*)$ and hence clearly

$$\| e^{At}x \|^2 \leq e^{\lambda t}\, \| x \|^2$$

or, $\|\cdot\|$, denoting operator norm of a matrix,

$$\|e^{At}\|^2 \leq e^{\lambda t}$$

and hence

$$\|e^{As} BB^* e^{A^*s}\| \leq \|BB^*\| \; \|e^{As}\|^2 \leq \|BB^*\| \; e^{\lambda s},$$

so that (6.37) converges as $t$ goes to infinity. Actually, we can replace this requirement [of 'stability' of the A-matrix] by a weaker condition that

$$[(A+A^*)x, x] \leq -|\lambda| \, [x, x] \quad \text{for } x \in \bigcup_t (\text{Range } e^{At}B)$$

but since this is quite transparent and only adds to the notation, we shall forego this, and stick to requiring that $A$ be stable. A related question is whether the limit, denoted $R(\infty)$ is non-singular; or, equivalently, for stable $A$, whether $(A, B)$ is 'controllable', that is to say

$$B, AB, \ldots A^{m-1} B$$

are linearly independent (as linear operators), $m$ being the dimension of $A$. Suppose now we assume $(A, B)$ is controllable so that $R(\infty)$ is non-singular. Then from (2.13) we have

$$0 = A\,R(\infty) + R(\infty)\,A^* + B\,B^*$$

so that also:

$$0 = KA + A^*K + K\,BB^*\,K \tag{6.38}$$

where

$$K^{-1} = R(\infty) \tag{6.}$$

Next, we shall confine ourselves to the case where 'signal and noise' are independent, namely:

$$B\,D^* = 0 \tag{6.}$$

Now from

$$R(t) = P(t) + \hat{R}(t)$$

it follows that $R(t)$ is non-singular as soon as $P(t)$ is. Conversely, suppose $P(t)$ is singular. Say:

$$v^* P(t)\, v = 0 \qquad \text{for some } m \times 1 \text{ vector } v.$$

This implies that

$$v^* x(t;\omega) = v^* \hat{x}(t;\omega)$$

But left side

$$= \int_0^t v^* e^{A(t-s)} B \, dW(s;\omega)$$

and right side

$$= \int_0^t v^* k(t,s)\; C\; x(s)\, ds + \int_0^t v^* k(t,s)\, D\, dW(s;\omega) \tag{6.}$$

where the first term:

$$= \int_0^t \int_\sigma^t v^* k(t, s) \ C \ e^{As} ds \ e^{-A\sigma} B \ dW(\sigma;\omega)$$

Since

$$B D^* = 0$$

it follows that we must have

$$v^* k(t, s) D = 0 \qquad 0 < s < t$$

or,

$$v^* k(t, s) D D^* = 0$$

or since $DD^*$ is non-singular,

$$v^* k(t, s) = 0 \qquad 0 < s < t$$

Hence both terms in (6.41) are zero. Hence so must

$$v^* e^{A(t-s)} B = 0 \qquad 0 < s < t \tag{6.42}$$

But this implies that $(A, B)$ is not controllable. Also from (6.42) it follows that

$$v^* R(t) = 0$$

But since $R(s)$ is monotone increasing, we must have

$$v^* R(s) = 0, \qquad 0 < s < t$$

Hence also

$$v^* P(s) = 0 \qquad 0 < s < t.$$

Hence $R(t)$ and $P(t)$ are singular or non-singular together. Note also that $R(t)$ is non-singular for any $t > 0$, implies that $(A, B)$ is controllable. Next let us note that $P(t)$ satisfies:

$$\dot{P}(t) = A\,P(t) + P(t)\,A^* + BB^* - P(t)\,C^*(DD^*)^{-1}C\,P(t)$$

In what follows we may set

$$DD^* = I$$

without loss of generality (simply redefine $C$ by $(\sqrt{DD^*})^{-1}\,C$)

Hence we write

$$\dot{P}(t) = A\,P(t) + P(t)\,A^* + BB^* - P(t)\,C^*C\,P(t) \qquad (6.4$$

Note that we can rewrite this as

$$\dot{P}(t) = (A-P(t)C^*C)\,P(t) + P(t)\,(A-P(t)C^*C) + BB^* + P(t)\,C^*C\,P(t) \qquad (6.4$$

Before we consider the asymptotic properties, we shall indicate a constructive method for solving (6.43), based essentially on Wonham [9 ].

An obvious iteration based on (6.44) would be

$$\dot{P}_{n+1}(t) = (A-P_n(t)C^*C)P_{n+1}(t) + P_{n+1}(t)\,(A-P_n(t)C^*C)$$

$$+ BB^* + P_n(t)\,C^*C\,P_n(t) \qquad (6.$$

and we see that $P_{n+1}$ is non-negative definite if $P_n$ is.

[If we based the iteration on (6.43), we would have

$$P_{n+1}(t) = R(t) - \int_0^t e^{A(t-s)}(P_n(s)\, C^*C\, P_n(s))\, e^{A^*(t-s)} ds \tag{6.46}$$

and here it does not follow that $P_{n+1}(t)$ is non-negative.]

We shall show that (6.45) is actually the Newton-Raphson 'equation - solving' algorithm for solving the 'equation':

$$P(t) = R(t) - \int_0^t e^{A(t-\sigma)}\, P(\sigma)\, C^*C\, P(\sigma)\, e^{A^*(t-\sigma)} d\sigma, \quad 0 \le t < \infty \tag{6.47}$$

Let C denote the Banach space of symmetric-matrix functions $P(\cdot)$ such that they are continuous on the closed interval $[0, \infty]$ [i.e., approach limits at $+\infty$], and vanishing at the origin. Then

$$Q(P) = q \qquad q(t) = \int_0^t e^{A(t-\sigma)} P(\sigma)\; C^*C\; P(\sigma)\; e^{A^*(t-\sigma)} d\sigma$$

clearly maps C into C. Moreover if we denote the Frechet derivative of $Q(\cdot)$ 'at' $P(\cdot)$ by $\mathscr{L}(P)$, we have

$$\mathscr{L}(P)\, f = g; \quad g(t) = \int_0^t e^{A(t-\sigma)}\, [f(\sigma) C^*C\; P(\sigma)\; C^*C\; f(\sigma)] d^{A^*(t-\sigma)} d\sigma$$

Note that

$$\mathscr{L}(P) = 2\; Q(P)$$

and that (6.47) can be expressed as a functional equation in $C$:

$$P+Q(P) - R = 0$$

The Newton-Raphson algorithm is given by:

$$P_{n+1} = P_n - (I + \mathscr{L}(P_n))^{-1} (P_n + Q(P_n) - R) \tag{6.4}$$

where $P_n$ stands for the function $P_n(t)$, $R$ for the function $R(t)$. Note that $\mathscr{L}(P)$ is a Volterra operator and $(I+ \mathscr{L}(P))$ has a bounded inverse. We shall now calculate the inverse in a form suitable for our purposes. Let

$$(I + \mathscr{L}(P))\, g = f$$

Then we have that $(f(t)-g(\cdot))$ is absolutely continuous and

$$\frac{d}{dt}(f(t)-g(t)) = A(f(t)-g(t) + (f(t)-g(t))A^* + g(t)\, C^*C\, P(t) + P(t)C^*C\, g(t)$$

Let us assume $f(t)$ is continuously differentiable. Then so is $g(t)$, and

$$\frac{d}{dt} g(t) = \frac{df}{dt}(t) - A\, f(t) - f(t)A^* - g(t)\, C^*C\, P(t) - P(t)\, C^*C\, g(t)$$

$$+ A\, g(t) + g(t)\, A^* \tag{6}$$

Hence if $f(\cdot)$ is continuously differentiable, we can determine $g(t)$, i.e., $(I + \mathscr{L}(P))^{-1} f$. Let us use this in (6.48). First we can rewrite (6.48) in the form:

$$P_n - P_{n+1} + 2\, Q(P_n) - \mathscr{L}(P_n)P_{n+1} = P_n + Q(P_n) - R$$

or, $P_{n+1}$ is defined by:

$$(I + \mathscr{L}(P_n))\, P_{n+1} = Q(P_n) + R$$

Here the right-side is indeed continuously differentiable. Hence, superdots indicate derivatives with respect to $t$, we have, using (6.49):

$$\dot{P}_{n+1} = \dot{Q}(P_n) + \dot{R} - A(Q(P_n) + R) - (Q(P_n) + R)A^* - P_{n+1}\, C^*C\, P_n$$

$$- P_n\, C^*C\, P_{n+1} + A\, P_{n+1} + A\, P_{n+1} + P_{n+1}\, A^*$$

But

$$\dot{Q}(P_n) = A\, Q(P_n) + Q(P_n)A^* + P_n C^*CP_n$$

$$\dot{R}(t) = AR(t) + R(t)A^* + BB^*$$

Hence

$$\dot{P}_{n+1} = BB^* + P_n C^*CP_n - P_{n+1}C^*CP_n - P_n C^*CP_{n+1} + AP_{n+1} + P_{n+1}A^*$$

$$= (A - P_n C^*C)P_{n+1} + P_{n+1}(A - P_n C^*C)^* + BB^* + P_n C^*CP_n \tag{6.50}$$

which is recognized as (6.45).

Let us next prove convergence of this algorithm.

<u>Theorem 6.3:</u> Suppose $A$ is stable. Suppose further that $C, A$ is observable that is to say that the matrix:

$$\int_0^\infty e^{A^*t}\, C^*C\, e^{At}\, dt$$

is nonsingular. Denote its inverse by $\Lambda$. Then the iteration (6.50) with $P_0(t)$ defined by:

$$P_0(t) = \int_0^t e^{As} \Lambda C^*C \Lambda e^{A^*s}\, ds$$

yields a sequence of (real) symmetric non-negative definite matrix functions in the space $C[0, \infty]$, vanishing at the origin. This sequence is moreover monotone:

$$P_{n+1}(t) \leq P_n(t)$$

and converges to $P(t)$, the unique solution of (6.43) in any finite interval of the form $[0, T]$. Moreover $P(t)$ converges as t goes to infinity, and denoting the limit by $P(\infty)$, we have:

$$0 = A P(\infty) + P(\infty)A^* + BB^* - P(\infty)C^*CP(\infty) \tag{6.51}$$

Further

$$A - P(\infty)C^*C$$

is stable (has all eigenvalues with strictly negative real parts). Finally, $P(\infty)$ is non-singular if and only if $R(\infty)$ is (or A–B is controllable).

<u>Proof</u> First let us note that

$$0 = A^*(\Lambda^{-1}) + (\Lambda^{-1}) A + C^*C$$

and hence:

$$0 = A\Lambda + \Lambda A^* + \Lambda C^*C \Lambda \tag{6.52}$$

and hence: $P_0(t)$ converges as t goes to infinity, and

$$P_0(\infty) = \int_0^\infty e^{At} \Lambda C^*C \Lambda e^{A^*t} \, dt = \Lambda$$

Next we need

Lemma 1: Let K be a non-negative definite real symmetric matrix such that

$$0 \geq (A-KC^*C)K + K(A-KC^*C)^* + \gamma BB^* + \mu\, KC^*CK \tag{6.53}$$

where

$\gamma \geq 0$; $\mu > 0$ (strictly positive!)

Then the eigenvalues of $(A-KC^*C)$ have strictly negative real parts.

Proof Let H denote $(A-KC^*C)^*$. Suppose for some nonzero vector z,

$$Hz = \lambda z \;\; ; \quad \text{Real part } \lambda = \sigma \geq 0$$

Then substituting in (6.53), we have

$$0 \geq 2\sigma \;\; [Kz, z] + \gamma \| B^* z \|^2 + \mu \| CKz \|^2$$

and hence

$$CKz = 0$$

which implies that

$$Hz = A^*z$$

and hence

$$[(A + A^*)\, z, z] = [(H + H^*)z, z] = 2\sigma[z, z]$$

so that $\sigma$ must be strictly negative.

Lemma 2. (Wonham [9]): Let $K, P$ denote two real symmetric, non-negative definite matrices and let

$$\Psi(K;P) = (A-KC^*C)P + P(A-KC^*C)^* + BB^* + KC^*CK$$

Then

$$\Psi(K;P) \geq \Psi(P;P) \tag{6.54}$$

Proof We have only to note that we can write:

$$\Psi(K;P) = \Psi(P;P) + (P-K)C^*CP + PC^*C(P-K) - PC^*CP + KC^*CP$$

$$= \Psi(P;P) + (K-P)C^*C(K-P)$$

$$\geq \Psi(P;P)$$

as required.

Lemma 3. Suppose $P_n(t)$ is a real symmetric non-negative definite matrix function uniformly continuous on $[0, \infty]$, with $P_n(0)$ equal to zero. Further suppose

$$(A - P_n(\infty)C^*C)$$

is stable.

Define $P_{n+1}(t)$ by:

$$\dot{P}_{n+1}(t) = (A-P_n(t)C^*C)P_{n+1}(t) + P_{n+1}(t)(A-P_n(t)C^*C)^* + BB^*$$

$$+P_n(t)C^*CP_n(t) \tag{6.55}$$

$$P_{n+1}(0) = 0$$



Then $P_{n+1}(t)$ has the same properties as $P_n(t)$.

Proof Since we are given that $P_n(t)$ converges as $t$ goes to infinity, and that

$$A-P_n(\infty)C^*C$$

is stable, it follows that for all $t > T$, $T$ sufficiently large, the eigenvalues of

$$A-P_n(t)C^*C$$

have also all strictly negative real parts, say all less than equal to $\sigma$, where $\sigma$ is negative. Hence if $\phi(t)$ denotes a fundamental matrix solution of:

$$\dot{\phi}(t) = (A-P_n(t)C^*C)\,\phi(t)$$

we have that $t > s > T$:

$$\|\phi(t)\phi(s)^{-1}x\| \leq \|x\| \exp \sigma(t-s) \tag{6.56}$$

Next let us note that we can express the solution of (6.55) as:

$$P_{n+1}(t) = \int_0^t \phi(t)\phi(s)^{-1}\,\theta(s)\,\phi(s)^{*-1}\phi(t)^*ds \tag{6.57}$$

where

$$\theta(s) = BB^* + P_n(s)C^*CP_n(s)$$

and the main thing to note is that $\theta(s)$ is convergent at infinity. From (6.57) it is immediate that $P_{n+1}(t)$ is non-negative definite. Let $\epsilon > 0$ be given. Then we can find $T$ large enough so that (6.56) holds and in addition

$$\|(\theta(t_2)-\theta(t_1))x\| \leq \epsilon\|x\|, \quad t_1, t_2 \geq T$$

Next let us note that for $\Delta$ sufficiently large so that, $\Delta \geq T$ and

$$\|(P_n(t)C^*C - P_n(\infty)C^*C)x\| \leq \epsilon\|x\|, \quad t \geq \Delta$$

we have

$$\|\phi(s+\Delta)\,\phi(\Delta)^{-1}x - e^{(A-P_n(\infty)C^*C)s}x\| \leq \epsilon\,\|x\| \tag{6.58}$$

For, setting

$$\dot{x}(t) = (A-P_n(t)C^*C)x(t); \; x(\Delta) = x$$

we have on the one hand

$$x(\Delta+s) = \phi(\Delta+s)\phi(\Delta)^{-1}x$$

while also:

$$x(\Delta+s) = \int_{\Delta}^{\Delta+s} e^{(A-P_n(\infty)C^*C)(\Delta+s-t)}(P_n(\infty)C^*C - P_n(t)C^*C)x(t)dt$$

$$+ e^{(A-P_n(\infty)C^*C)s}x$$

A simple estimation of the integral using (6.56) verifies (6.58); then we can write:

$$P_{n+1}(t) = \int_0^t \phi(t)\phi(t-s)^{-1}\,\theta(t-s)\,\phi(t-s)^{*-1}\,\phi(t)^* ds$$

and hence:

$$P_{n+1}(t_2) - P_{n+1}(t_1) = \int_0^T e^{(A-P_n(\infty)C^*C)s}\,(\theta(t_2-s)$$

$$- \theta(t_1-s))e^{(A-P_n(\infty)C^*C)^* s}\,ds$$

+ terms which go to zero with $T \to \infty$ by virtue of our estimates as can be directly verified.

The first term, for $t_2 \geq t_1 \geq 2\Delta$, is less than (in norm)

$$\epsilon/2|\sigma|$$

Hence $P_{n+1}(t)$ converges as $t$ goes to infinity. Hence from (6.54), $\dot{P}_{n+1}(t)$ also converges, and hence must have zero for its limit. Hence we have (in the notation of Lemma 2):

$$0 = \Psi(P_n(\infty);\, P_{n+1}(\infty)) \geq \Psi(P_{n+1}(\infty);\, P_{n+1}(\infty)) \tag{6.59}$$

by Lemma 2; and the last inequality, by Lemma 1 implies that

$$A-P_{n+1}(\infty)C^*C$$

is stable. This completes proof of the Lemma.

Next let us note that $P_0(t)$ satisfies the conditions of Lemma 3. For, from (6.52) we have:

$$0 = (A-\Lambda C^*C)\Lambda + \Lambda(A-\Lambda C^*C)^* + 3\Lambda C^*CC$$

and by Lemma 1, this implies the stability of $(A-\Lambda C^*C)$. Of course $P_0(t)$ converges as $t \to \infty$. Hence (6.50) yields the kind of sequence asserted. The monotonicity follows from Lemma 2. Thus following Wonham as in [ 9 ], we have:

$$\dot{P}_n(t) - \Psi(P_n(t);P_n(t)) \geq \dot{P}_n(t) - \Psi(P_{n-1}(t);P_n(t))$$

$$= 0$$

$$= \dot{P}_{n+1}(t) - \Psi(P_n(t);P_{n+1}(t))$$

so that

$$\dot{P}_n(t) - \dot{P}_{n+1}(t) \geq \Psi(P_n(t);P_n(t)) - \Psi(P_n(t);P_{n+1}(t))$$

$$= H_n(t)(P_n(t)-P_{n+1}(t)) + (P_n(t) - P_{n+1}(t))H_n(t)^*$$

But if

$$\dot{Z}(t) \geq H_n(t)Z(t) + Z(t)H_n(t)^*$$

we must have:

$$\frac{d}{dt}(\phi(t)^{-1}Z(t)\,\phi(t)^{*-1}) \geq 0$$

as can be verified by calculation. Since $Z(0)$ is zero, it follows that

$$Z(t) \geq 0$$

Hence $P_n(t)$ is monotone non-increasing. In particular we have

$$P_{n+1}(\infty) \leq P_n(\infty)$$

Because $P_n(t)$ is non-negative definite, and the sequence $P_n(t)$ is monotone, we have that $P_n(t)$ converges for every $t$ in the closed interval $[0, \infty]$. From (6.55), $\dot{P}_n(t)$ also converges, and it is evident that, denoting the limit of $P_n(t)$ by $P(t)$, $\dot{P}(t)$, $\dot{P}_n(t)$ converges to $\dot{P}(t)$, and $P(t)$ is of course the unique solution of (6.42). We have thus obtained a constructive method for solving (6.42); particularly noteworthy is the monotonic nature of the approximating sequence. Let us now examine the asymptotic properties.

Lemma 4: Under the condition that $A$ is stable, (6.51) has a unique solution, in the class of real symmetric matrices.

Proof Let $P, Q$, denote two real symmetric matrix solutions of (6.51). First of all, by Lemma 1, both $(A-PC^*C)$ and $(A-QC^*C)$ are stable. Substituting $P, Q$ into equation (6.51) we have, upon subtraction:

$$A(P-Q) + (P-Q)A^* + QC^*CQ - PC^*CP = 0$$

Let $z$ be an eigen-vector of $(P-Q)$:

$$(P-Q)z = \lambda z, \ \lambda \text{ must of course be real.}$$

Then we have:

$$\lambda [(A+A^*)z, z] + \|CQz\|^2 - \|CPz\|^2 = 0$$

But

$$\|CQz\|^2 - \|CPz\|^2 = [C(P+Q)z, C(Q-P)z]$$

$$= (-\lambda) [C^*C(P+Q)z, z]$$

Hence if $\lambda$ is not zero,

$$[(A+A^*)z, z] - [C^*C(P+Q)z, z] = 0 = [(A-C^*CP)z, z] + [(A^*-QC^*C)z, z]$$

which contradicts the stability condition. Hence all eigen-values must be zero, or P must equal Q.

Next let us note that we have also a constructive method for solving (6.51). Thus we have (see (6.59)):

$$0 = (A-P_n(\infty)C^*C)P_{n+1}(\infty) + P_{n+1}(\infty)(A-P_n(\infty)C^*C)^* + BB^* + P_n(\infty)C^*CP_n(\infty)$$

which is a linear equation for determining $P_{n+1}(\infty)$ from $P_n(\infty)$, and taking

$$P_0(\infty) = \Lambda$$

we have further that

$$P_{n+1}(\infty) \leq P_n(\infty)$$

so that $P_n(\infty)$ converges monotonically to the solution of (6.51).

Finally, that $P(t)$ converges to $P(\infty)$ as $t$ goes to infinity follows from:

Lemma 5. The solution $P(t)$ of (6.45) with $P(0) = 0$, is actually monotonic non-decreasing as $t$ increases.

Proof We follow Wonham [ 7 ]. Thus let (in the notation of Lemma 2)

$$\dot{\tilde{P}}(t) = \Psi(P(t+\tau); \tilde{P}(t)) \; ; \; \tilde{P}(0) = 0 \; ; \quad \tau > 0 \quad \text{and fixed.}$$

Then from Lemma 2, we have:

$$\dot{P}(t) \leq \Psi(P(t+\tau); P(t))$$

and hence

$$\dot{\tilde{P}}(t) - \dot{P}(t) \geq \Psi(P(t+\tau); \tilde{P}(t)) - \Psi(P(t+\tau); P(t)))$$

which, just as in the proof of the monotonicity of the sequence $P_n(\cdot)$ in Lemma 3, implies that

$$\tilde{P}(t) \geq P(t)$$

But

$$\tilde{P}(t) = \int_0^t \phi(t+\tau)\, \phi(s+\tau)^{-1}(BB^* + P(s+\tau)C^*CP(s+\tau))\phi(s+\tau)^{*-1}\phi(t+\tau)^* ds$$

and by an obvious change of variable in the integrand, this is

$$= \int_{\tau}^{t+\tau} \phi(t+\tau)\phi(s)^{-1}(BB^* + P(s)C^*CP(s))\phi(s)^{*-1}\phi(t+\tau)^* ds$$

$$= P(t+\tau) - \int_{0}^{\tau} \phi(t+\tau)\phi(s)^{-1}(BB^* + P(s)C^*CP(s))\phi(s)^{*-1}\phi(t+\tau)^* ds$$

$$\leq P(t+\tau)$$

Hence

$P(t) \leq P(t+\tau)$ as required.

Hence $P(t)$ converges as $t$ goes to infinity to the unique solution of (6.51). Finally suppose $P(\infty)$ is singular. Then by Lemma 5, so is $P(t)$ for every $t$, and as we have seen, this implies that (A-B) is not controllable.

# CHAPTER VII

## LINEAR STOCHASTIC CONTROL

Let us next consider stochastic 'control' problems for the linear system:

$$x(t;\omega) = \int_0^t A(s)\, x(s;\omega)ds + \int_0^t B(s)\, u(s)\, ds + \int_0^t F(s)\, dW(s;\omega),$$

$$0 \leq t \leq 1 \tag{7.1}$$

$$Y(t;\omega) = \int_0^t C(s)\, x(s;\omega)ds + \int_0^t G(s)\, dW(s;\omega), \quad 0 \leq t \leq 1$$

where $u(.)$ denotes the 'control' function to be determined, and where we assume that all the coefficients are continuous on $[0,1]$, and

$$G(s)G(s)^* > 0 \qquad 0 \leq s \leq 1$$

and of course $W(s;\omega)$ is a Wiener process as before. For a general survey of such problems, see [7]. The first control problem we shall consider is the following. The control $u(t)$ at time $t$ must "depend" only on the "observation" up to time $t$ that is on $Y(s;\omega)$, $0 < s < t$. We wish to minimize:

$$\int_0^1 E\big([Q(s)x(s;\omega),\ x(s;\omega)]\big)\, ds + \lambda \int_0^1 E[u(t),\ u(t)]\, dt \tag{7.2}$$

where $Q(s)$ is continuous in $s$ and is non-negative definite, and $\lambda$ is a fixed positive constant. Since $u(t)$ is now also a random process, let us denote it by $u(t;\omega)$. It is implicit that $u(t;\omega)$ is jointly measurable in $t$ and $\omega$, and obviously we can assume:

$$\int_0^1 E[u(t;\omega),\ u(t;\omega)\,] < \infty$$

For practical application of this problem, which is perhaps the most used part of the theory, see [22, 38, 41].

Let H denote the class of controls of the desired kind: i.e.

i) $\int_0^1 E\|u(t;\omega)\|^2 dt < \infty$ ii) $u(t;\omega)$ measurable $\beta_Y(t)$ for each t.

Whatever the choice of $u(t;\omega)$ in H, let

$$x_u(t;\omega) = \int_0^t A(s)x_u(s;\omega)\,ds + \int_0^t B(s)u(s;\omega)\,ds$$

$$\tilde{Y}(t;\omega) = Y(t;\omega) - Y_u(t;\omega);\; Y_u(t;\omega) = \int_0^t C(s)x_u(s;\omega)\,ds$$

and let $\tilde{x}(t;\omega) = x(t;\omega) - x_u(t;\omega)$ ; $\hat{\tilde{x}}(t;\omega) = E[\tilde{x}(t;\omega) \mid \beta_{\tilde{Y}}(t)]$

$$\hat{x}(t;\omega) = \hat{\tilde{x}}(t;\omega) + x_u(t;\omega)$$

$\beta_{\tilde{Y}}(t)$ denoting as before the sigma-algebra generated by $\tilde{Y}(s;\omega)$, $0 \leq s \leq t$.
Then we have the Kalman filter equations characterizing $\hat{\tilde{x}}(t;\omega)$:

$$Z_0(t;\omega) = \tilde{Y}(t;\omega) - \int_0^t C(\sigma)\hat{\tilde{x}}(\sigma;\omega)\,d\sigma \quad ; \quad \beta_{\tilde{Y}}(t) = \beta_{Z_0}(t);$$

$$\hat{\tilde{x}}(t;\omega) = \int_0^t A(s)\hat{\tilde{x}}(s;\omega)\,ds + \int_0^t (P(s)C(s)^* + F(s)G(s)^*)(G(s)G(s)^*)^{-1} dZ_0(s;\omega)$$

And hence, $\hat{x}(t;\omega)$ is the unique solution of

$$\hat{x}(t;\omega) = \int_0^t A(s)\hat{x}(s;\omega) + \int_0^t (P(s)C(s)^* + F(s)G(s)^*)(G(s)G(s)^*)^{-1} dZ_0(s;\omega)$$
$$+ \int_0^t B(s)u(s;\omega)\,ds \tag{7.3}$$

and since

$$Y(s;\omega) - \int_0^s C(\sigma)\hat{x}(\sigma;\omega)\,d\sigma = \tilde{Y}(s;\omega) - \int_0^t C(\sigma)\hat{\tilde{x}}(\sigma;\omega)\,d\sigma\,,$$

it is important to note that

$$Z_0(s;\omega) = Y(s;\omega) - \int_0^s C(\sigma)\hat{x}(\sigma;\omega)\,d\sigma \tag{7.4}$$

is a Gaussian Martingale with covariance $G(s)G(s)^*$. (Cf. (6.30), (6.12)).

The matrix $P(s)$ is determined by: (Cf. (6.31)):

$$P(t) = E[e(t;\omega)e(t;\omega)^*]; \quad e(t;\omega) = x(t;\omega) - \hat{x}(t;\omega) = \tilde{x}(t;\omega) - \hat{\tilde{x}}(t;\omega)$$

$$\dot{P}(t) = A(t)P(t) + P(t)A(t)^* + F(t)F(t)^* - (P(t)C(t)^* + F(t)G(t)^*) \cdot$$

$$(G(t)G(t)^*)^{-1} (C(t)P(t) + G(t)F(t)^*) \tag{7.5}$$

and in particular does not depend on the control.

Since $u(t;\omega)$ depends only on the observation, it is reasonable to conclude that we need only consider controls such that

$$E[x_u(s;\omega)\; e(s;\omega)^*] = 0 \tag{7.6}$$

This would imply in particular that

$$E[Q(s)\; x(s;\omega),\; x(s;\omega)]$$

$$= E[Q(s)\hat{x}(s;\omega),\; \hat{x}(s;\omega)] + \text{Tr. } Q(s)\; P(s) \tag{7.7}$$

But (7.6) requires a formal proof, and before we get into that, let us exploit first our knowledge of deterministic control theory. Thus, let us fix the sample point $\omega$, and consider the problem of minimizing:

$$\int_0^1 [Q(s)x(s;\omega),\; x(s;\omega)]ds + \lambda \int_0^1 [u(s;\omega),\; u(s;\omega)]\; ds \tag{7.8}$$

for each $\omega$. For this purpose it is convenient to write

$$v(t;\omega) = \int_0^t (P(s)C(s)^* + F(s)G(s)^*)(G(s)G(s)^*)^{-1} \, dZ_0(s;\omega)$$

$$w(t;\omega) = \int_0^t \phi(t)\,\phi(s)^{-1} \, d\, v(s;\omega) + e(t;\omega)$$

where $\phi(t)$ is a fundamental matrix solution of

$$\dot{\phi}(t) = A(t)\,\phi(t)$$

and finally:

$$x(t;\omega) = \int_0^t \phi(t)\phi(s)^{-1} B(s)\, u(s;\omega)\, ds + w(t;\omega) \tag{7.}$$

Note that $w(t;\omega)$ is continuous in $t$(as we have seen). Since $\omega$ is fixed, we consider the control functions as functions of $t$ alone for the moment. Let $L_2(0,1)$ denote the usual $L_2$ space for control functions $u(t)$. Introduce the linear bounded operator on this space:

$$Lf = g\ ;\quad g(t) = \int_0^t \phi(t)\phi(s)^{-1} B(s)\, f(s) ds, \quad 0 \leq t \leq 1$$

Then, we can rewrite (7.8) as:

$$[Q(Lu + w),\ (Lu + w)] + \lambda\, [u, u] \tag{7.}$$

where $Q$ stands for the operator corresponding to multiplication by $Q(s)$, $u$ stands for $u(t)$, $w$ for $w(t;\omega)$, and inner-products in two spaces have been used. Being a quadratic form with $\lambda$ positive, it is obvious by a routine first variation (gradient with respect to $u$) that the unique minimum is given by

$$\lambda u_0 + L^*Q L u_0 = - L^*Q w \tag{7.11}$$

where $L^*$ denotes the adjoint of $L$, and is given by:

$$L^*f = h \; ; \; h(t) = \int_t^1 B(t)^* \phi(t)^{*-1} \phi(s)^* f(s)ds \quad 0 \le t \le 1$$

From (7.11) we have:

$$u_0(t;\omega) = -B(t)^* z(t;\omega)/\lambda \tag{7.12}$$

where

$$z(t;\omega) = \int_t^1 \phi(t)^{*-1} \phi(s)^* Q(s)x(s;\omega)ds \tag{7.13}$$

and is the unique solution of:

$$\dot{z}(t;\omega) = -A(t)^* z(t;\omega) - Q(t) x(t;\omega) \; ; \; z(1;\omega) = 0 \tag{7.14}$$

Here (7.12) is the best 'open loop' solution. It is unfortunately NOT acceptable since it does not meet the requirement of being dependent only on the observation up to time t - it is NOT, in other words, measurable $B_Y(t)$, the sigma algebra generated by $Y(s;\omega)$, $0 \le s \le t$.

Next we observe that the equation

$$\dot{P}(t) + P(t)A(t) + A(t)^* P(t) + Q(t) - P(t)B(t)B(t)^*P(t)/\lambda = 0; \; P(1) = 0 \tag{7.15}$$

has a unique (symmetric) solution in $0 \le t \le 1$, which we shall denote by $P_c(t)$. We shall now show that we have the decomposition:

$$u_0(t;\omega) = u_1(t;\omega) + u_2(t;\omega)$$

where

$$u_1(t;\omega) = -B(t)^* P_c(t)\, \hat{x}(t;\omega)/\lambda \tag{7.16}$$

$$u_2(t;\omega) = \frac{+B(t)^*}{\lambda} \int_t^1 \Psi(t)^{*-1}\, \Psi(s)^* \,(P_c(s)dv(s;\omega) + Q(s)e(s;\omega)ds). \tag{7.17}$$

where

$$\dot{\Psi}(t) = \left(A(t) - \frac{B(t)B(t)^*P_c(t)}{\lambda}\right)\Psi(t)$$

This follows from the directly verifiable relation (using (7.4), (7.13), (7.14), and (7.15)) in differential form:

$$d(z(t;\omega) - P_c(t)\,\hat{x}(t;\omega)) =$$

$$= -\left(A(t)^* - \frac{P_c(t)B(t)B(t)^*}{\lambda}\right)(z(t;\omega) - P_c(t)\hat{x}(t;\omega))dt - P_c(t)dv(t;\omega) - Q(t)e(t;\omega)dt$$

and since

$$z(1;\omega) = 0; \quad P_c(1) = 0$$

We observe now that $u_2(t;\omega)$ depends on the "future", and heuristically, then, we should expect this component is not "available" to the observer and that the optimal control based on available data upto time $t$ should be $u_1(t;\omega)$. Moreover, if we define the control by (7.16), then substituting into (7.3), we see that $\hat{x}(t;\omega)$ is measurable $\beta_{Z_0}(t)$ $(= \beta_{\tilde{Y}}(t))$, and then so is $u_1(t;\omega)$, so that from

$$Y(t;\omega) = \tilde{Y}(t;\omega) + \int_0^t C(s)\, x_u(s;\omega)\, ds$$

we have that:

$$\beta_Y(t) = \beta_{Z_0}(t) = \beta_{\tilde{Y}}(t) \qquad 0 \le t < 1$$

From this in turn, we have that

$$\hat{x}(t;\omega) = E(x(t;\omega) \mid \beta_Y(t))$$

and hence (7.6) follows also.

Let us now try to derive this rigorously. First we prove:

Theorem 7.1 Let $H_u$ denote the class of controls in H with the property that

$$\beta_Y(t) = \beta_{Z_0}(t) \qquad 0 \leq t \leq 1$$

Then $H_u$ is a Hilbert space with inner-product defined by:

$$[u, q] = \int_0^1 E[u(t;\omega), q(t;\omega)]\, dt$$

The space $H_u$ contains functions of any one of the following forms:

$$u(t;\omega) = \int_0^t k(t;s) dY(s;\omega) \tag{7.18a}$$

$$u(t;\omega) = \int_0^t k(t;s) dZ_o(s;\omega) \tag{7.18b}$$

$$u(t;\omega) = \int_0^t k(t;s)\, \hat{x}(s;\omega) ds \tag{7.18c}$$

where k(t; s) is square integrable on the triangle : $0 \leq s \leq t \leq 1$. Further, any one of the forms implies the others.

Proof It is immediate that $H_u$ is a Hilbert space. Let k(t;s) be any (Lebesgue measurable) square integrable function on the unit square. Let us define $u(t;\omega)$ by:

$$u(t;\omega) = \int_0^t k(t;s)\, dZ_o(s;\omega)$$

Then we shall show that it is measurable with respect to $\beta_Y(t)$. For this, let us note that by substituting $u(t;\omega)$ into (7.3), we have:

$$\hat{x}(t;\omega) = \int_0^t \Psi(t;s)\, dZ_o(s;\omega)$$

where $\Psi(t, s)$ is square integrable on the unit square. Substituting this into

$$Y(t;\omega) = Z_o(t;\omega) + \int_0^t C(s)\,\hat{x}(s;\omega)\,ds$$

we have:

$$Y(t;\omega) = Z_o(t;\omega) + \int_0^t J(t;s)\,dZ_o(s;\omega)$$

where $J(t,s)$ is given by:

$$J(t;s) = \int_s^t C(\sigma)\,\Psi(\sigma;s)\,d\sigma$$

Hence, just as in the proof of Theorem 6.1, we have that

$$\beta_{Z_o}(t) = \beta_Y(t)$$

for each t. Hence $u(t;\omega)$ is measurable $\beta_Y(t)$ as required. Further

$$\int_0^1 E(\|u(t;\omega)\|^2)dt = \int_0^1 \int_0^t \text{Tr. } k(t;s)G(s)G(s)^*k(t;s)^*ds\,dt$$

and it readily follows that the subspace of elements of the form (7.18b) is actually closed. Next let us note that if (7.18a) holds we have also the representation:

$$u(t;\omega) = \int_0^t h(t;s)\,dZ_o(s;\omega)$$

which follows from the fact that using (7.4), we have

$$\hat{x}(t;\omega) = \int_0^t m(t;s)\,dY(s;\omega)$$

where $m(t;s)$ is square integrable on the unit square and hence also:

$$Z_o(t;\omega) = Y(t;\omega) - \int_0^t r(t;s)dY(s;\omega);$$

$$r(t;s) = \int_s^t C(\sigma)\, m(\sigma;s)\, d\sigma$$

so that as in the proof of Theorem 6.1, we have again that

$$\beta_Y(t) = \mathcal{B}_{Z_o}(t)$$

Finally if we have the representation (7.18c), substituting into (7.3) we see that it has a unique solution and hence readily obtain the representation (7.18b). We have only exhibited linear transformations on the data as members of $H_u$. But it contains many non-linear transformations; for example let $\phi(t;x)$ be any function satisfying Lipschitz condition that:

$$\| \phi(t;x) - \phi(t;y) \| \leq \text{const.} \|x - y\| \quad 0 \leq t \leq 1$$

Then the equation (7.3) with $u(t;\omega) = \phi(x(t;\omega))$, which is now a non-linear Ito equation, has a unique solution measurable $\beta_{Z_o}(t)$, as we shall show in the second volume (see [3], [6] for a proof). So does the same equation with $dY(t;\omega)$ in place of $dZ_o(t;\omega)$ using the transformation (7.4) ; which shows that $\hat{x}(t;\omega)$ is also measurable $\beta_Y(t)$.

As a final point, let it be noted that for any $u(t;\omega)$ in $H_u$ we have that

$$\hat{x}(t;\omega) = E(x(t;\omega)/\beta_Y(t)).$$

We shall now prove that the control defined by (7.16) is optimal in $H_u$. Thus, we prove optimality in the class of linear transformations on the observation, non-linear transformations being outside the scope of the present volume. It may nevertheless be of interest to note here that despite much evidence, it is not known yet whether $H_u$ is equal to $H$;

the argument by Bucy (22, p.100) is inadequate, and the problem is the subject of much current research.[+]

Next the linear transformation L on $H_u$ defined by

$$Lu = v; \quad v(t;\omega) = \int_0^t \phi(t)\,\phi(s)^{-1}\,B(s)\,u(s;\omega)\,ds$$

is bounded; let us denote the adjoint (with respect to $H_u$) by $L^*$. Then we have:

$$[Lu, y] = \int_0^1 E([u(t;\omega),\ q(t;\omega)])\,dt$$

where

$$q(t;\omega) = B(t)^* \int_t^1 \phi(t)^{*-1}\,\phi(s)^*\,y(s;\omega)\,ds$$

Now $u(t;\omega)$ is given to be measurable $\beta_{Z_o}(t)$. Hence if

$$r(t;\omega) = E\left(q(t;\omega)/\beta_{Z_o}(t)\right)$$

we have that

$$[Lu, y] = [u, r]$$

If $r(t;\omega)$ is of one of the forms (7.18b) or (7.18c), then it is in $H_u$. This is the way we shall calculate $L^*$ in what follows. Next, it is

---

[+] See U.V. Rozanov: Paper presented at the Conference on Multivariate Analysis, Dayton, Ohio, June, 1972. See also [37].

clear that the functional (7.2) can be cast into the form:

$$[Q(Lu + w), Lu + w] + \lambda[u, u]$$

with $w(t;\omega)$ as in (7.9), and

yields a quadratic form over $H_u$. And hence the minimum is attained at the unique point $u_o(t;\omega)$ in $H_u$, and further

$$u_o = (-1/\lambda)L^*(Q(Lu_o + w)) \quad =$$

Of course $(L u_o + w)$ is the function $x(t;\omega)$, and since $u_o(t;\omega)$ and $w(t;\omega)$ are measurable $\beta_{Z_o}(t)$, it follows that so is $x(t;\omega)$. We shall now show that

$$L^*Q(Lu + w)$$

is given by

$$B(t)^* \, E\left(\int_t^1 \phi(t)^{*-1} \phi(s)^* \, Q(s) \, x(s;\omega) ds \,\Big|\, \beta_{Z_o}(t)\right) \tag{7.19}$$

by showing that it is of the form (7.18b). First of all, (7.19) is given by

$$B(t)^* \hat{z}(t;\omega)$$

where

$$\hat{z}(t;\omega) = E(z(t;\omega)/\beta_{Z_o}(t))$$

where $z(t;\omega)$ is the unique solution of:

$$\dot{z}(t;\omega) = -A(t)^* z(t;\omega) - Q(t) x(t;\omega) \; ; \quad z(1;\omega) = 0$$

next let us note that

$$u_o = (L^*QL + \lambda I)^{-1} L^*Qw$$

Since $w$ has the form:

$$w(t;\omega) = \int_0^t h(t;s)\, dZ_o(s;\omega) + e(t;\omega)$$

and

$$E[e(t+\Delta;\omega) | \beta_{Z_o}(t)] = 0, \text{ for } \Delta \geq 0$$

it follows that

$$B(t)^* E(\int_t^1 \phi(t)^{*-1} \phi(s)^* Q(s) w(s;\omega)ds / \beta_{Z_o}(t))$$

is of the form (7.18b), and hence is the element $L^*Qw$. Let $\mathscr{L}_u$ denote the subspace of elements of the form (7.1 b). Then $\mathscr{L}_u$ is closed, and $L^*QL$ maps $\mathscr{L}_u$ into itself. Hence it follows that

$$(L^*QL + \lambda I)^{-1}, \quad \lambda > 0$$

maps $\mathscr{L}_u$ into itself, and that $u_o$ belongs to $\mathscr{L}_u$. Hence $\hat{x}(t;\omega)$ has the form:

$$\hat{x}(t;\omega) = \int_0^t k(t;s)\, dZ_o(s;\omega)$$

and hence it follows that (7.19) is also in $\mathscr{L}_u$, and thus is equal to

$$L^*Q(Lu + w)$$

In particular then we have that

$$u_o(t;\omega) = (-1/\lambda) B(t)^* \hat{z}(t;\omega)$$

**Then we have, in differential notation:**

$$d(z(t;\omega) - P_c(t)\hat{x}(t;\omega)) = -(A^* - P_c BB^*/\lambda)(z(t;\omega) - P_c(t)\hat{x}(t;\omega))\, dt$$

$$-(P_c BB^*/\lambda)(z(t;\omega). - \hat{z}(t;\omega))dt - P_c(t)dv(t;\omega)$$

(7

Because $z(1;\omega)$ and $P_c(1)$ vanish and

$$E((z(s;\omega) - \hat{z}(s;\omega)) \mid \beta_{Z_o}(t)) = 0 \quad \text{for} \quad s \geq t$$

it follows that:

$$E(z(t;\omega) - P_c(t)\hat{x}(t;\omega)) / \beta_{Z_o}(t)) = 0$$

thus proving the optimality of

$$\frac{-B(t)^*}{\lambda} P_c(t) \hat{x}(t;\omega)$$

Note that the filtering and control can thus be treated separately - this is referred to as the 'separation' principle - see Wonham [7]. It was first derived by Joseph and Tou [12]. Our treatment is quite different from both of these.

It should be noted that $P_c(t)$ is independent of the state noise variance, that is, Tr. $F(s)F(s)^*$.

Finally let us remove the objection that (7.20) was brought in ad hoc. The dependence on this equation can be eliminated in the following way. Starting with

$$\hat{x}(t;\omega) = \int_0^t k(t;s)\, dZ_o(s;\omega)$$

we observe that $\hat{x}(t;\omega)$ is a Gaussian process, and if it were Markov, equation (7.19) would simplify. Moreover, since $u_o$ is unique, if after the assumption of Markov property we can still satisfy

$$u_o = (L^*QL + \lambda I)^{-1} L^* w$$

we know we will have the optimal solution. Now, if $\hat{x}(t;\omega)$ were Markov, we must have

$$E(\hat{x}(s;\omega)/\beta_{Z_o}(t)\,, \quad t \leq s) = M(s)\, M(t)^{-1}\, \hat{x}(t;\omega) \tag{7}$$

where $M(t)$ is a fundamental matrix solution of

$$\dot{M}(t) = H(t)M(t)$$

and

$$\hat{x}(t;\omega) = \int_0^t H(s)\, \hat{x}(s;\omega)\, ds + v(t;\omega)$$

Substituting (7.21) into (7.19) we obtain:

$$u_o(t;\omega) = (-1/\lambda)\, B(t)^*\, K(t)\, \hat{x}(t;\omega) \tag{7.2}$$

where

$$K(t) = \int_t^1 \phi(t)^{*-1}\, \phi(s)^*\, Q(s)\, M(s)\, M(t)^{-1}\, ds$$

We note that $K(t)$ satisfies:

$$\dot{K}(t) + A(t)^*\, K(t) + Q(t) + K(t)\, H(t) \quad ; \quad K(1) = 0 \tag{7.2}$$

But using the expression (7.22) for $u_o(t;\omega)$, we must have:

$$H(t) = (A(t) - (1/\lambda)\, B(t)\, B(t)^*\, K(t))$$

and hence (7.23) is thus the Riccati equation for $K(t) = P_c(t)$.

Steady State Control: Time Invariant Systems

If the system is time-invariant, that is to say, if the matrices $A, B, C, D, F, G, Q$ are all independent of $t$, then often a more useful case in practice is one where the upperlimit in time is not fixed, and we invoke time-averages instead. More specifically, we now seek a control $u(t;\omega)$ measurable $B_Y(t)$, so as to minimize:

$$\lim_{T\to\infty} \left\{ (1/T) \int_0^T [Qx(t;\omega), x(t;\omega)] \, dt + (1/T) \int_0^T [u(t;\omega), u(t;\omega)] \, dt \right\} \tag{7.24}$$

excepting possibly an $\omega$-set of measure zero.
We assume now that $A$ is stable; and for simplicity that

$$FG^* = 0$$

In addition we assume "Observability":

$$\int_0^\infty e^{A^*t} C^*(GG^*)^{-1} C e^{At} dt \qquad \text{is nonsingular}$$

as well as 'controllability':

$$\int_0^\infty e^{At} BB^* e^{A^*t} dt \qquad \text{is nonsingular}$$

The observability ensures a unique non-negative definite matrix solution of:

$$AP + PA^* + FF^* - PC^*(GG^*)^{-1} CP = 0 \tag{7.25}$$

which we denote by $P_f$ (f standing for filter). Similarly controllability ensures a unique solution of

$$PA + A^*P + Q - P B B^*P = 0 \tag{7.26}$$

which we denote by $P_c$ (c for control). We note that $P_f$ is the limit of $P(t)$ defined by (7.5) as t goes to infinity.

It is implicit in (7.24) that the limits exist, and since we are minimizing it is enough to consider the controls for which the limits are finite. Actually we shall need to consider even a smaller class of controls. Thus we shall require that the controls are such that:

$$\lim (1/T) \int_0^T x(t;\omega)\, x(t+s;\omega)^*\, dt = \lim_{t\to\infty} E(x(t;\omega)x(t+s;\omega)^*) = R(s) \tag{7.27}$$

with probability one for every s. To ensure this, we now specify the class of controls to be such that

(i) $u(t;\omega)$ jointly measurable in $t$ and $\omega$, and measurable $B_Y(t)$ for each $t$; and $B_Y(t) = B_{Z_o}(t)$

(ii) $\lim (1/T) \int_0^T u(t;\omega)u(t+s;\omega)^*)\, dt = \lim_{t\to\infty} E(u(t;\omega)u(t+s;\omega)^*) = R_u(s)$

(iii) $\lim (1/T) \int_0^T u(t;\omega)\, w(t;\omega)^* dt = \lim_{t\to\infty} E(u(t;\omega)w(t;\omega)^*)$

the equality holding with probability one in (ii) and (iii) and the limits finite. We now consider the space of control functions with properties above and further such that for any two such functions $u_1(t;\omega)$, $u_2(t;\omega)$, we have:

$$\text{(iv)} \quad \lim_{T\to\infty} (1/T) \int_0^T u_1(t;\omega)\, u_2(t+s;\omega)^* dt = \lim_{t\to\infty} E(u_1(t;\omega)u_2(t+s;\omega)^*) = R_{12}(s)$$

the limit finite, and equality holding with probability one; and define $\text{Tr. } R_{12}(0)$ to be the inner-product. It is easy to see that the space is non-trivial. Thus let $k(t)$, $t \geq 0$, be any function continuous on finite intervals and such that

$$\int_0^\infty \text{Tr. } k(t)\, GG^*\, k(t)^*\, dt < \infty \tag{7.28}$$

and define

$$u(t;\omega) = \int_0^t k(t-s)\, dZ_0(s;\omega) \tag{7.29}$$

Then it is easy to see that $u(t;\omega)$ satisfies (i), (ii) and (iii), and that the class of such functions is a Hilbert space. Further for any two functions $k_1(t)$, $k_2(t)$ satisfying (7.28), the corresponding controls defined by (7.29) satisfy (iv). Since the class of functions satisfying (7.28) is a linear class, we see that we do have a Hilbert space which is non-trivial. In other words there is a Hilbert space of control functions satisfying (i) through (iv) containing control functions of the form (7.29). Let us denote this Hilbert space by $H_u$. Note that for controls of the form (7.29), we have:

$$\|u\|^2 = \int_0^\infty \text{Tr. } k(t)\, G\, G^*\, k(t)^*\, dt$$

and that $w(t;\omega)$ is also in $H_u$.

Next let us define the linear bounded transformation $L$ on $H_u$

$$Lu = q\ ;\quad q(t;\omega) = \int_0^t e^{A(t-s)}\, B\, u(s;\omega)\, ds \tag{7.30}$$

Then $q(t;\omega)$ has properties (i) through (iii), and hence

$$\hat{x}(t;\omega) = q(t;\omega) + w(t;\omega)$$

has property (7.27), and similarly, so does $x(t;\omega)$. Moreover:

$$\lim_{T\to\infty} (1/T) \int_0^T ([Qx(t;\omega),\ x(t;\omega)] - [Q\hat{x}(t;\omega),\ \hat{x}(t;\omega)])\, dt$$

$$= \lim_{t\to\infty} \text{Tr. } Q(R(t) - E(\hat{x}(t;\omega)\hat{x}(t;\omega)^*))$$

$$= \lim_{t\to\infty} \text{Tr. } QP(t)$$

$$= \text{Tr. } QP_f$$

Hence it is enough to minimize:

$$\lim_{T\to\infty} (1/T)\int_0^T [Q\,\hat{x}(t;\omega), \hat{x}(t;\omega)]\,dt + \lambda \lim_{T\to\infty} \frac{1}{T}\int_0^T [u(t;\omega), u(t;\omega)]\,dt$$

$$= [Q(Lu + w), Lu + w] + \lambda\,[u, u] \qquad (7.$$

where $L$ maps $H_u$ into a similar Hilbert space of functions $x(t;\omega)$, $y(t;\omega)$ with inner product given by:

$$\lim_{T\to\infty} \frac{1}{T}\int_0^T [x(t;\omega), y(t;\omega)]\,dt = \lim_{t\to\infty} E([x(t;\omega), y(t;\omega)])$$

The adjoint mapping $L^*$ is defined by:

$$L^* y = u\ ;\ u(t;\omega) = E\left(\int_t^\infty B^* e^{A^*(s-t)}\, y(s;\omega)\, ds \,\middle|\, B_Y(t)\right)$$

This follows readily from the fact that

$$\lim_{T\to\infty} (1/T)\int_0^T \left[\int_0^t e^{A(t-s)} B\, u(s;\omega) ds,\ y(t;\omega)\right] dt$$

$$= \lim\ (1/T)\int_0^T \left[u(s;\omega),\ \int_0^\infty B^* e^{A^*(t-s)}\, y(t;\omega)\, dt\right] ds$$

$$= \lim\ (1/T)\int_0^T \left[u(s;\omega),\ E\left(\int_s^\infty B^* e^{A^*(t-s)} y(t;\omega)\, dt \,\middle|\, B_Y(s)\right)\right] ds$$

$$+ \lim\ (1/T)\int_0^T [u(s;\omega),\ g(s;\omega)]\, ds$$

where

$$g(s;\omega) = \int_s^\infty B^* e^{A^*(t-s)} (y(t;\omega) - E(y(t;\omega)/B_Y(s))dt$$

and

$$\lim (1/T) \int_0^T [u(s;\omega), g(s;\omega)]ds = \lim_{s \to \infty} E([u(s;\omega), g(s;\omega)])$$

$$= 0, \text{ since } E([u(s;\omega), g(s;\omega)]) = 0 \text{ for every } s.$$

As before then, the optimal control $u_0(t;\omega)$ is thus given by:

$$u_o = (-1/\lambda)L^* (Lu_o + w)$$

Or,

$$u_o(s;\omega) = (-B^*/\lambda) E(\int_s^\infty e^{A^*(t-s)} Q \hat{x}(t;\omega) dt / B_Y(s))$$

Let us use the notation:

$$z(s;\omega) = \int_s^\infty e^{A^*(t-s)} Q \hat{x}(t;\omega) dt \tag{7.32}$$

$$\hat{z}(s;\omega) = E(z(s;\omega)/B_Y(s))$$

so that

$$\dot{z}(s;\omega) + A^* z(s;\omega) + Q \hat{x}(s;\omega) = 0 \tag{7.33}$$

$$u_0(t;\omega) = (-B^*/\lambda) \hat{z}(t;\omega)$$

Moreover, analogous to (7.20), we have using (7.26), (7.32), in differential form:

$$d(z(t;\omega) - P_c\hat{x}(t;\omega)) = (-1)(A^* - P_c BB^*/\lambda)(z(t;\omega) - P_c\hat{x}(t;\omega))dt$$

$$-P_c(BB^*/\lambda)(z(t;\omega) - \hat{z}(t;\omega))dt - P_c dv(t;\omega)$$

For convenience let us use the abbreviated notation:

$$r(t;\omega) = z(t;\omega) - P_c\hat{x}(t;\omega) \; ; \; H = (A^* - P_c BB^*/\lambda)(-1)$$

Let $t$ be fixed and let $T > t$. Then we have:

$$r(t;\omega) = e^{H(t-T)} r(T;\omega) + \int_t^T e^{H(s-t)}(P_c(BB^*/\lambda)(z(s;\omega) - \hat{z}(s;\omega))ds + P_c dv(s;\omega))$$

and hence it follows that

$$E(r(t;\omega)/B_Y(t)) = e^{H(t-T)} E(r(T;\omega)/B_Y(t)) \quad (7.34)$$

Now as we know, the eigenvalues of $H$ have all positive real parts. Hence letting $T$ go to infinity in (7.34), and noting that

$$E(\|E(r(T;\omega)/B_Y(t))\|^2) \leq E(\|r(T;\omega)\|^2)$$

and hence we have that, since the right-side is bounded from above,

$$E(\|E(r(t;\omega)/B_Y(t))\|^2) = 0$$

Hence,

$$\hat{z}(t;\omega) - P_c\hat{x}(t;\omega) = 0$$

Or,

$$u_o(t;\omega) = -(B^*/\lambda)P_c\hat{x}(t;\omega)$$

## Final Value Problems

Another commonly used cost function involves the 'terminal' or final time (the so-called "Final Value" Problem): (using $T$ to denote the final time in (7.1) instead of 1): we seek to minimize:

$$E([Qx(T;\omega), x(T;\omega)]) + \lambda \int_0^T E([u(t;\omega), u(t;\omega)])\, dt \tag{7.35}$$

over controls $u(t;\omega)$ in $H$. For the same reasons as before we shall again obtain the optimal control in $H_u$ only. Define the linear transformation $L$ mapping $H_u$ into a Hilbert space of random variables measurable $B_Y(T)$, by:

$$Lu = y \ ; \ y(\omega) = \int_0^T \Phi(T)\, \Phi(s)^{-1}\, B(s)\, u(s;\omega) ds \ ; \ \dot{\Phi}(t) = A(t)\, \Phi(t)$$

Denoting the adjoint by $L^*$, we have:

$$L^*y = u \ ; \ u(t;\omega) = E(B(t)^*\, \Phi(t)^{*-1}\, \Phi(T)\, y(\omega)/B_{Z_o}(t))$$

Proceeding as before it is clear that the optimal control denoted $u_o(t;\omega)$ is given by:

$$u_o = \frac{(-1)}{\lambda}\, L^*\, Q(Lu + w)$$

it follows as before that

$$\hat{z}(t;\omega) = P_c(t)\hat{x}(t;\omega)$$

'Tracking' Problem

A useful generalization of the cost function is to include a "tracking" feature. Thus, suppose $j(t;\omega)$ is jointly measurable in $t$ and $\omega$, and we want to minimize (analogous to (7.2)), over $H_u$, the functional:

$$\int_0^1 E(\|J(t)x(t;\omega) - j(t;\omega)\|^2)\,dt + \int_0^1 E([u(t;\omega),\ u(t;\omega)])\,dt$$

where, say $J(t)$ is continuous in $t$, and

$$\int_0^1 E(\|j(t;\omega)\|^2)dt < \infty\ .$$

Now:

$$E(\|J(t)x(t;\omega) - j(t;\omega)\|^2) = E(\|J(t)(x(t;\omega) - \hat{x}(t;\omega))\|^2) + E[\|j(t;\omega)\|^2]$$

$$+ E(\|J(t)\hat{x}(t;\omega)\|^2 - 2E([J(t)\hat{x}(t;\omega), j(t;\omega)])$$

$$- 2E([J(t)(x(t;\omega) - \hat{x}(t;\omega)),\ j(t;\omega)])$$

Since the last term does NOT involve the control, we are left with the problem of minimizing:

$$\int_0^1 \Big\{ E([Q(t)\hat{x}(t;\omega),\ \hat{x}(t;\omega)])dt - 2\ E([\hat{x}(t;\omega),\ k(t;\omega)])dt$$

$$+ \lambda E([u(t;\omega),\ u(t;\omega)])dt \Big\}$$

where

$$Q(T) = J(t)^* J(t)$$

$$k(t;\omega) = J(t)^* j(t;\omega)$$

It is readily seen that the optimal control is again given by:

$$u_0(t;\omega) = -(1/\lambda)\, B(t)^* \,(P_c(t)\, \hat{x}(t;\omega) - \hat{m}(t;\omega))$$

where

$$\hat{m}(t;\omega) = E(m(t;\omega)/B_Y(t)) \quad = \quad E[m(t;\omega) \,|\, B_{Z_o}(t)]$$

$$\dot{m}(t;\omega) + A(t)^* \, m(t;\omega) = k(t;\omega) \;;\; m(1;\omega) = 0$$

and $P_c(t)$ is again the solution of (7.15). Note in particular that if $j(t;\omega)$ is deterministic, then $\hat{m}(t;\omega) = m(t;\omega)$. The extensions to the other cases clearly go in analogous fashion.

## Differential Games with Imperfect Information

As a final application, we shall consider a class of stochastic differential games for linear systems. Specifically, we shall study a zero-sum two-person game with the same data available to both players with a fixed time quadratic cost functional. Thus using the subscript p to denote one of the players referred to as the pursuer, and the subscript e to indicate the other player referred to as the evader, we have the following problem: Find:

$$\underset{u_e}{\text{Sup}} \quad \underset{u_p}{\text{Inf}} \quad E([Q\, x(T;\omega),\, x(T;\omega)])$$

$$+ \lambda \int_0^T E[\|u_p(t;\omega)\|]^2 - \mu \int_0^T E[\|u_e(t;\omega)\|^2]\, dt; \tag{7.38}$$

$$\lambda > 0, \quad \mu > 0$$

where the state equations are : (analagous to (7.1):

$$x(t;\omega) = \int_0^t A(s)x(s;\omega)ds + \int_0^t B_p(s)u_p(s;\omega)ds + \int_0^t B_e(s)u_e(s;\omega)ds + \int_0^t F(s)dW(s;\omega) \;;\; 0 \le t \le T$$

and the observation is:

$$Y(t;\omega) = \int_0^t C(s)x(s;\omega)\, ds + \int_0^t G(s)dW(s;\omega) \tag{7.39}$$

As before we make the blanket assumption that all coefficient functions are continuous. It is required that the controls $u_p(t;\omega)$, $u_e(t;\omega)$ be jointly measurable in $t$ and $\omega$, measurable $B_Y(t)$ for each $t$. It is convenient to use customary terminology and refer to such controls as admissible. Let $H_p$ denote the Hilbert space as before of pursuer controls analagous to $H_u$ and similarly $H_e$ the Hilbert space of evader controls, analagous to $H_u$, and recall that we assume $B_Y(t) = B_{Z_o}(t)$. Define the functional on $H_p \times H_e$.

$$c(u_p;u_e) = E([Qx(T;\omega), x(T;\omega)]) + \lambda \|u_p\|^2 - \mu \|u_e\|^2 \tag{7.4}$$

This is clearly a continuous functional, in fact quadratic. The game is said to have a value if

$$\operatorname*{Sup}_{u_e \in H_e} \operatorname*{Inf}_{u_p \in H_p} c(u_p;u_e) = \operatorname*{Inf}_{u_p \in H_p} \operatorname*{Sup}_{u_e \in H_e} c(u_p;u_e)$$

and the corresponding number is defined to be the value. We shall denote it by $c$. A pair of admissible pursuer-evader controls is said to be a 'saddle point' if

$$c(u_{op};u_e) \le c(u_{op};u_{oe}) = c \le c(u_p;u_{oe}) \tag{7.4}$$

and the corresponding controls $u_{op}, u_{oe}$ are referred to as optimal strategies or optimal controls. The following theorem characterizing optimal strategies is an immediate extension of our results on Final Value Control.

Theorem 7.2

For the functional (7.38) there exist two constants $\mu_1 > \mu_o \geq 0$ such that for $\mu \leq \mu_o$ the value of the game is infinite while for $\mu_o < \mu \leq \mu_1$ the game has no value. For $\mu > \mu_1$, the game has a finite value and a unique saddle point exists. Moreover it is given by:

$$u_{op}(t;\omega) = -(1/\lambda)B_p^*(t)\, P_g(t)\hat{x}(t;\omega) \tag{7.42}$$

$$u_{oe}(t;\omega) = (+1/\mu)B_e^*(t)P_g(t)\hat{x}(t;\omega)$$

where $P_g(t)$ satisfies:

$$\dot{P}_g(t) + P_g(t)A(t) + A^*(t) + P_g(t)\left[\frac{B_e(t)B_e(t)^*}{\mu} - \frac{B_p(t)B_p(t)^*}{\lambda}\right]P_g(t)$$

$$= 0;\ P_g(T) = Q$$

Of course

$$\hat{x}(t;\omega) = E(x(t;\omega)/B_{Z_o}(t)) = E(x(t;\omega)\big| B_Y(t)) \tag{7.43}$$

and satisfies (7.3) with $B_p(t)\, u_p(t) + B_e(t)\, u_e(t)$ in place of $B(t)u(t)$.

Proof

Let $H_p$ and $H_e$ denote the same spaces as before. Then with $\hat{x}(t;\omega)$ as before (satisfying (7.43) we have again

$$E([Qx(T;\omega), x(T;\omega)]) = E([Q\hat{x}(T;\omega), \hat{x}(T;\omega)]) + \text{Tr } QP(T)$$

Let us now introduce the operators $L_p$, $L_e$ by:

$$L_p u_p = g; \quad g(T;\omega) = \int_0^T \Phi(T)\Phi(s)^{-1} B_p(s) u_p(s;\omega) ds \tag{7.44}$$

$$L_e u_e = f; \quad f(T;\omega) = \int_0^T \Phi(T)\Phi(s)^{-1} B_e(s) u_e(s;\omega) ds \tag{7.45}$$

with $w$ as before, we need only to deal with the functional:

$$f(u_p;u_e) = \|L_p u_p + L_e u_e + w\|^2 + \lambda \|u_p\|^2 - \mu \|u_e\|^2 \tag{7.46}$$

where we have not bothered to distinguish the norms by their corresponding spaces, being self-evident. Note that the functional (7.46) is not necessarily concave in $u_e$. It is certainly convex in $u_p$. For fixed $u_e$, let us first minimize with respect to $u_p$. We know that the minimum is attained at the unique point:

$$(-1)(L_p^* L_p + \lambda I)^{-1} \; L_p^*(L_e u_e + w) \tag{7.47}$$

and the actual minimal value itself being:

$$(-1)[(L_p^* L_p + \lambda I)^{-1} \; L_p^*(L_e u_e + w), \; L_p^*(L_e u_e + w)] - \mu \|u_e\|^2 + \|L_e u_e + w\|^2 \tag{7.48}$$

The main thing to be noted is that this minimal value is actually independent of $u_p$. Moreover, we observe (7.48) is a quadratic form in $u_e$. Let us denote it by $g(u_e)$. Then we can rewrite it as:

$$g(u_e) = [L_e^* (I-R) L_e u_e, u_e] - \mu[u_e, u_e] + 2[u_e, z] + [(I-R)w, w]$$

where

$$R = L_p(L^*L_p + \lambda I)^{-1} L_p^* \ ; \quad z = L_e^*(I-R)w$$

Let

$$\operatorname*{Inf}_{u_e} \ [L_e^*(I-R)L_e u_e, u_e] / [u_e, u_e] = \mu_o$$

Then $\mu_o$ is actually non-negative since the numerator is equal to the sum of the first two terms in (7.46) for some $u_p$, with w set equal to zero. Hence for $\mu \leq \mu_o$, the sum of the first two terms in $g(u_e)$ will be non-negative, so that

$$\operatorname*{Sup}_{u_e} \ g(u_e) = \infty$$

And since

$$f(u_p;u_e) \geq g(u_e)$$

it follows that

$$\operatorname*{Sup}_{u_e} \ \operatorname*{Inf}_{u_p} \ f(u_p;u_e) = +\infty = \operatorname*{Inf}_{u_p} \ \operatorname*{Sup}_{u_e} \ f(u_p;u_e) \ , \qquad \mu \leq \mu_o .$$

Moreover, for $\mu > \mu_o$, $g(u_e)$ is concave in $u_e$, and further the maximum is attained at a unique point which we denote by $u_{oe}$. Hence

$$\operatorname*{Sup}_{u_e} \ \operatorname*{Inf}_{u_p} \ f(u_p;u_e) = g(u_{oe}) < \infty \quad \text{for } \mu > \mu_o .$$

Next let

$$\mu_1 = \operatorname*{Sup}_{u_e} \ \|L_e u_e\|^2 / \|u_e\|^2$$

Then for $\mu \leq \mu_1$, $f(u_p;u_e)$ is convex in $u_e$, and hence

$$\sup_{u_e} f(u_p;u_e) = +\infty$$

so that the game has no value. On the other hand for $\mu > \mu_1$, the functional $f(u_p;u_e)$ is convex in $u_p$ and concave in $u_e$ and hence by the usual theory [13], the game has a finite value attained by a saddle point, which in our case is unique, and is in fact clearly given by $u_{op}, u_{oe}$, where $u_{op}$ is given by (7.47) with $u_e = u_{oe}$. Then

$$u_{op} = (-1/\lambda)L_p^* Q \hat{x}(T;\omega)$$

$$u_{oe} = (+1/\mu)L_e^* Q \hat{x}(T;\omega)$$

from which the remainder of statements of the theorem follow as in the Final Value problem. Thus we have:

$$u_{op} = (-1/\lambda)\ B_p(t)^* \ \hat{z}(t;\omega)$$

$$u_{oe} = \frac{1}{\mu} B_e(t)^* \ \hat{z}(t;\omega)$$

$$\hat{z}(t;\omega) = E[\ z(t;\omega) | B_{Z_0}(t)]$$

$$\dot{z}(t;\omega) + A^*(t)\ z(t;\omega) = 0; \quad z(T;\omega) = Q\ \hat{x}(T;\omega)$$

$$d(z(t;\omega) - P_g(t)\ \hat{x}(t;\omega)\ )$$

$$= (-1)\left(A^*(t) - \frac{P_g(t)B_p(t)B_p(t)^*}{\lambda} + \frac{P_g(t)B_e(t)B_e(t)^*}{\mu}\right)(z(t;\omega) - P_g(t)\ \hat{x}(t;\omega))dt$$

$$+ P_g(t)\left(\frac{B_p(t)B_p^*(t)}{\lambda} - \frac{B_e(t)B_e(t)^*}{\mu}\right)(\hat{z}(t;\omega) - z(t;\omega))dt$$

$$+ P_g(t)\ dv(t;\omega)$$

and as before, since

$$z(T;\omega) - P_g(T)\,\hat{x}(T;\omega) = 0$$

this implies that

$$\hat{z}(t;\omega) = P_g(t)\,\hat{x}(t;\omega)$$

## CHAPTER VIII

## SYSTEM IDENTIFICATION

In this final chapter, we consider the problem of System Identification† which (largely because of the increasing use of high-speed large-memory digital computers) has been growing interest in recent years. As we have seen, before we can apply the theories of stochastic control, it is necessary to know the parameters characterizing the system (that is, the matrices A, B, F, etc.). In many cases, in a sense in all cases, these are not known in sufficient precision and have to be deduced from measurements made while the system is operating. The measurements are subject to error; for many purposes the errors can be modelled as additive Gaussian noise. The particular problem we shall consider here is that of identifying a linear dynamic system driven by state 'noise' as well as known inputs, from observed output in additive white noise. (See [41] for a direct application ).

More precisely, let

$$x(t;\omega) = \int_0^t A\, x(s;\omega)ds + \int_0^t B\, u(s)ds + \int_0^t F\, dW(s;\omega) \tag{8.1}$$

$$y(t;\omega) = \int_0^t C\, x(s;\omega)\, ds + \int_0^t D\, u(s)ds + \int_0^t G\, dW(s;\omega) \tag{8.2}$$

-------------

†An excellent survey of the general area is given in [42]

where $W(s;\omega)$ is a Wiener process, $u(s)$ is a known (given) input, and

$$FG^* = 0 \quad \text{(state noise is independent of observation noise).}$$

The identification problem is that of estimating unknown parameters in the matrices $A, B, F, C, D$ and $G$, given

$$y(t;\omega), \quad 0 < t < T$$

$$u(t), \quad 0 < t < T$$

In our approach to this problem we assume that there exists a 'true' set of the unknown parameters which we shall indicate by the subscript o. We shall return to the question of uniqueness presently. Let us use

$\theta$ to denote the vector of unknown parameters with $\theta_o$ denoting the true value. An estimate based on measurements over a time-interval $T$ will be denoted $\theta_T(\omega)$. We shall say that a system is identifiable if we can find ('compute') $\theta_T(\omega)$ such that

(i) $E(\theta_T(\omega)) \rightarrow \theta_o$ as $T$ goes to infinity.

(ii) $\lim \theta_T(\omega) = \theta_o$ in probability.

In other words, we require an asymptotically unbiassed, consistent estimate. Our main result will be to show that under certain sufficiency conditions ('identifiability conditions') it is possible to find such estimates in a constructive way.

We begin with a discussion of the basic estimation problem; specifically we shall only consider the 'maximum likelihood' estimate.

Let us recall that in the usual case of finite dimensional observation, if x denotes the observed vector variable whose distribution is absolutely continuous with respect to Lebesque measure, with density $p(x/\theta)$, then the maximum likelihood estimate is the value of $\theta$ which maximizes $p(x/\theta)$ for the given x. From the computational point of view however we can only find 'local' maxima characterized by the vanishing of the gradient. Hence if we know that there is a neighborhood of $\theta_o$ in which $\theta_o$ is the only local maximum and if we can produce a computational algorithm which is shown to converge provided we start from a point in this neighborhood, we have arrived at a constructive method for finding the estimate. This is the approach we take here; making necessary allowances for the non-finite dimensional nature of the observation space, and the asymptotics.

Let us fix T and first calculate the likelihood functional. We shall only consider the case where the matrix G is known completely and

$$GG^* > 0 \tag{8.3}$$

Also we assume:

$$\sup_t \|u(t)\| < \infty$$

Then by specializing (6.36) we obtain:

Theorem 8.1 Let $y(t;\omega)$ be defined by equations (8.2) and (8.1). Let

$$y_1(t;\omega) = \sqrt{(GG^*)^{-1}} \quad y(t;\omega)$$

Then the measure induced by the process $y_1(t;\omega)$, $0 < t < T$, is absolutely continuous with respect to Wiener measure, and the likelihood functional is given by:

$$H_1(y_1(\cdot;\omega)) = H(y(\cdot;\omega))$$

$$= \exp - 1/2 \left\{ \int_0^T [(GG^*)^{-1} (D\,u(t) + C\hat{x}(t;\omega)),\ Du(t) + C\hat{x}(t;\omega)) \right.$$

$$\left. -2 \int_0^T [(GG^*)^{-1} (Du(t) + C\hat{x}(t;\omega)),\ dy(t;\omega)] \right\} \tag{8.4}$$

where

$$\hat{x}(t;\omega) = \int_0^t A\,\hat{x}(s;\omega)ds + \int_0^t B\,u(s)ds + \int_0^t P(s)C^*(GG^*)^{-1}\,dy(s;\omega)$$

$$- \int_0^t P(s)C^*(GG^*)^{-1}\,C\,\hat{x}(s;\omega)ds$$

$$- \int_0^t P(s)C^*(GG^*)^{-1}\,D\,u(s)ds \tag{8.5}$$

$$\dot{P}(t) = AP(t) + P(t)A^* + FF^* - P(t)C^*(GG^*)^{-1}\,CP(t);\ P(0) = 0 \tag{8.5a}$$

and

$$Z\ (t;\omega) = y(t;\omega) - \int_0^t C\,\hat{x}(s;\omega)ds - \int_0^t D\,u(s)ds \tag{8.6}$$

is such that

$$\sqrt{(GG^*)^{-1}}\ Z\ (t;\omega) \tag{8.7}$$

is a Wiener process.

Proof  We have only to note that

$$y_1(t;\omega) = \int_0^t C_1\,x(s;\omega)\,ds + \int_0^t G_1\,dW(s;\omega) + D_1 \int_0^t u(s)ds$$

where

$$C_1 = \sqrt{(GG^*)^{-1}}\,C \;;\quad G_1 = \sqrt{(GG^*)^{-1}}\,G \;;\quad D_1 = \sqrt{(GG^*)^{-1}}\;D$$

and substitute into (6.36) with

$$u(t) \sim B\,u(t)$$

$$v(t) \sim D_1 u(t)$$

therein.

Let us next note that (8. 4) necessarily defines the likelihood functional (R-N derivative) only for the true value of the parameters, $\theta = \theta_o$. Nevertheless, we can define (8.4) as a functional on the process for all values of the parameters, (i.e. for any $\theta$) with $\hat{x}(t;\omega)$ defined by (8.5) even though it does not have the interpretation as conditional expectation, and $Z(t;\omega)$ does not necessarily yield a Wiener process.

From now on we shall need to indicate the dependence on $\boldsymbol{\theta}$ explicitly. Thus, let $L(\theta)$ denote the Volterra operator:

$$L(\theta)f = g \;;\; g(t) = C\,\phi(t) \int_0^t \phi(s)^{-1}\; P(s)C^*(GG^*)^{-1}\; f(s)ds \tag{8.8}$$

where $\phi(t)$ is a fundamental matrix solution of:

$$\dot{\phi}(t) = (A - P(t)C^*(GG^*)^{-1}C))\phi(t) \tag{8.9}$$

Let

$$m(\theta;t) = D\,u(t) + \int_0^t C(\exp(A(t-s))\;B\,u(s)ds \tag{8.10}$$

To be explicit, let us now define $\hat{x}(\theta;t;\omega)$ by (repeating (8.5)):

$$\hat{x}(\theta;t;\omega) = A\int_0^t \hat{x}(\theta, s;\omega)\, ds + B\int_0^t u(s)\, ds + \int_0^t P(s)\, C^*\, (GG^*)^{-1} dy(s;\omega)$$

$$- \int_0^t P(s)\, C^*\, (GG^*)^{-1}(C\hat{x}(\theta;s;\omega) + Du(s))\, ds$$

and

$$Z(\theta;t;\omega) = y(t;\omega) - \int_0^t (C\hat{x}(\theta;s;\omega) + Du(s))\, ds \tag{8.11}$$

Then $Z(\theta_o;t;\omega)$ is a Wiener process; and defining

$$\tilde{y}(\theta;t;\omega) = y(t;\omega) - \int_0^t m(\theta;s;\omega)\, ds$$

(corresponding to subtracting the "suspected" response due to the known inputs), and defining similarly

$$\tilde{x}(\theta;t;\omega) = \hat{x}(\theta;t;\omega) - x_u(t)$$

where

$$x_u(t) = \int_0^t A\, x_u(s)\, ds + \int_0^t B\, u(s)\, ds$$

we note that, by subtracting the corresponding equations

$$\tilde{x}(\theta;t;\omega) = \int_0^t A\, \tilde{x}(\theta;s;\omega)\, ds + \int_0^t P(s)\, C^*(GG^*)^{-1}\, dZ(\theta;s;\omega) \tag{8.12}$$

On the other hand

$$\begin{aligned} Z(t;\theta;\omega) &= y(t;\omega) - \int_0^t C\, \hat{x}(\theta;s;\omega)\, ds - \int_0^t D\, u(s)ds \\ &= y(t;\omega) - \int_0^t C\, \tilde{x}(\theta;s;\omega)ds - \int_0^t m(\theta;s)ds \\ &= \tilde{y}(\theta;t;\omega) - \int_0^t C\, \tilde{x}(\theta;s;\omega)ds \end{aligned} \tag{8.13}$$

so that also:

$$\tilde{x}(\theta;t;\omega) = \int_0^t A\, \tilde{x}(\theta;s;\omega)ds \quad + \int_0^t P(s)\, C^*(GG^*)^{-1}(d\tilde{y}(\theta;t;\omega) - C\tilde{x}(\theta;s;\omega)ds) \tag{8.14}$$

Hence

$$C\hat{x}(\theta;t;\omega) + D\, u(t) = C\, \tilde{x}(\theta;t;\omega) + m(\theta;t)$$

$$= h(\theta;t;\omega) + m(\theta;t)$$

where

$$h(\theta;t;\omega) = \int_0^t C\, e^{A(t-s)}\, P(s)\, C^*(GG^*)^{-1}\, dZ(\theta;s;\omega) \tag{8.15}$$

$$= \int_0^t C\, \phi(t)\, \phi(s)^{-1}\, P(s)\, C^*(GG^*)^{-1}\, d\tilde{y}(\theta;s;\omega) \tag{8.16}$$

Introducing the operator $K(\theta)$ by:

$$K(\theta)f = g; \quad g(t) = \int_0^t C(\exp\,(t-s)A)\; P(s)C^*\,(GG^*)^{-1}\, f(s)\, ds$$

We can also see from (8.12), (8.13), (8.14), I denoting the identity operator:

$$K(\theta) = L(\theta)\,(I + K(\theta))$$

or,

$$(I + K(\theta))^{-1} = I - L(\theta)$$

Based on (8.4), we now define the functional $H(\theta;y(.;\omega);T)$ by:

$$H(\theta;y(\cdot;\omega);T) = \exp\cdot(-1/2)\left\{\int_0^T [(GG^*)^{-1}(h(\theta;t;\omega) + m(\theta;t)),\, h(\theta;t;\omega) + m(\theta;t)]\,dt - 2\int_0^T [(GG^*)^{-1}(h(\theta;t;\omega) + m(\theta;t)),\, dy(t;\omega)]\right\}. \tag{8.17}$$

Let

$$[f, g] = \int_0^T [(GG^*)^{-1}\, f(t),\; g(t)]dt\; ; \; \|f\|^2 = [f, f]$$

$$L(\theta)d\tilde{y}(\theta;\cdot;\omega) \sim h(\theta;.;\omega) \sim K(\theta)\, dZ(\theta;\cdot;\omega)$$

so that we can write (in short-hand form):

$$H(\theta;y(.;\omega);T) = \exp(-1/2)\Big\{\|m(\theta;.) + L(\theta)(dy(.;\omega) - m(\theta;.))\|^2$$

$$-2\,[m(\theta;.) + L(\theta)(dy(.;\omega) - m(\theta;.)), dy(.;\omega)]\Big\}$$

Let

$$q(\theta;y(.;\omega);T) = (-2/T)\,\mathrm{Log}\,H(\theta;y(.;\omega);T) \qquad (8.18)$$

Our basic technique will be to show that for sufficiently large T, the functional (8.18) ( the "log likelihood functional") will have a maximum at the true value of $\theta$; or, more accurately, that the gradient of (8.18) will have a root (that is, will be zero) at the true value $\theta = \theta_o$, and at no other value of $\theta$ in a sufficiently small neighbourhood about $\theta_o$. This last point is particularly important in any computational technique, in order that we do not converge to a "wrong" root. It is thus essential that we have the following condition hold:

Condition S: There is a neighborhood of $\theta_o$ such that no other value of $\theta$ will yield a response identical to the one observed for all $t > 0$, for any $\omega$, omitting only a set of zero probability.

There is one case where this condition does not obviously hold. Suppose that all the matrices A through F are unknown. Then this condition is NOT satisfied. For, let $S_n$, $n > 1$, be a sequence of matrices of the same dimensions as A such that $S_n$ converges to the identity matrix. Let $\theta_o \sim (A_o, B_o, F_o, D_o)$, and let

$$A_n = S_n A_o S_n^{-1}$$

$$B_n = S_n B_o$$

$$F_n = S_n F_o$$

$$C_n = C_o S_n^{-1}$$

Then it is readily verified that $\theta_n \sim A_n, B_n, C_n, D_o, F_n$ yields the same mean response $m(\theta_n;t)$ as at $\theta_o$, for all $\theta_n$, and that the covariance of the process $y(t;\omega)$ is the same. Clearly $\theta_n$ converges to $\theta_o$ and hence condition S is violated. [On the other hand, we note that $\theta_n$ all yield the same response and hence cannot be distinguished based on u(.) and $y(.;\omega)$ alone.] For S to hold, it is thus necessary that enough parameters of the matrices A, B, C, F are known. Fixing some parameters can be looked upon as a linear equality constraint in addition on $\theta$. It is easy to demonstrate that, with not all parameters unknown, the assumption S can indeed be satisfied. In fact, we shall eventually state a sufficient condition (S′) which can be verified and which implies S. In practical problems the unknown parameters usually have a physical significance which will enable us to assert that assumption S must hold. From now on, we shall consider only $\theta$ in such a neighborhood which we shall denote by $\mathcal{N}$. We clearly may, and do, take $\mathcal{N}$ to be bounded.

To motivate our method for estimating $\theta_o$, let us next examine some asymptotic properties of (8.18). In order that asymptotic limits exist, we assume, from now on, that for any point $\theta$ in the closure of $\mathcal{N}$ :

A is stable; C-A is observable (8.19)

Further, we shall assume that the input u(.) is such that:

$$\lim_{T\to\infty} (1/T) \int_0^T \|u(t)\|^2 \, dt < \infty \quad \text{(exists and is finite)} \tag{8.20}$$

$$r_u(t) = \lim_{T\to\infty} (1/T) \int_0^T u(s)\, u(s+t)^* \, ds$$

is a continuous function of t in every finite interval.

From (8.19) it follows that:

$$\lim_{t\to\infty} P(t) \quad \text{exists}$$

We shall denote the limit simply by P.

From (8.19) and (8.20) it follows that the "time-average"

$$\lim_{T\to\infty} (1/T) \int_0^T [m(\theta;t),\ m(\theta;t)]dt$$

also exists. We denote the limit by $|| m(\theta)||^2_{av.}$.

Next we note that in $\mathcal{N}$, $q(\theta;y(.;\omega);T)$, for fixed T and $\omega$, is actually "analytic" in $\theta$; that, in fact, (multiple) power series expansions are valid in the closure of $\mathcal{N}$. The gradient is denoted:

$$\nabla_\theta q(\theta;y(.;\omega);T)$$

and is of course a vector with components

$$q_i(\theta;y(.;\omega);T) = \frac{\partial}{\partial\theta_i} q(\theta;y(.;\omega);T)$$

"$\theta_i$" being the "components" of $\theta$.

Similarly, we use the notation:

$$q_{ij}(\theta;y(.;\omega);T) = \frac{\partial^2}{\partial\theta_i\partial\theta_j} q(\theta;y(.;\omega);T)$$

Next let

$$q(\theta;T) = E(q(\theta;y(.;\omega);T))$$

$$q_i(\theta;T) = E(\ q_i(\theta;y(.;\omega);T)$$

$$q_{ij}(\theta;T) = E(q_{ij}(\theta;y(.;\omega);T))$$

To calculate these expectations, it is convenient to rewrite $q(\theta;y(.;\omega);T)$ as:

$$\begin{aligned} q(\theta;y(.;\omega);T) = (1/T) \Big\{ &|| m(\theta;.) + L(\theta)((I + K(\theta_o))dZ(\theta_o;.;\omega) + m(\theta_o;.) \\ &- m(\theta;.))||^2 - 2\,[m(\theta;.) + L(\theta)((I+K(\theta_o))dZ(\theta_o;.;\omega) \\ &+ m(\theta_o;.) - m(\theta;.)),\quad (I+K(\theta_o))dZ(\theta_o;.;\omega) + m(\theta_o;.)\,]\Big\} \end{aligned} \quad (8.21)$$

substituting for $dy(t;\omega)$ the expression:

$$dy(t;\omega) = dZ(\theta_o;t;\omega) + (K(\theta_o)dZ(\theta_o;.;\omega)) + m(\theta_o;t))dt$$

so that we can take advantage of the fact that $Z(\theta_o;t;\omega)$ is a Wiener process with covariance $GG^*$. Thus a tedious but straight forward calculation using (8.21) yields:

$$\begin{aligned} q(\theta;T) = (1/T)\Big\{ & (\| m(\theta;.) + L(\theta)(m(\theta_o;.) - m(\theta;.)) \|^2 \\ & - 2[m(\theta;.) + L(\theta)(m(\theta_o;.) - m(\theta;.)), m(\theta_o;.)] \\ & + [L(\theta)(I+K(\theta_o)), L(\theta)(I+K(\theta_o))] - 2[K(\theta_o), L(\theta)(I+K(\theta_o))]\Big\} \end{aligned} \quad (8.22)$$

where for Volterra operators $K_1$, $K_2$ with kernels $K_1(t;s)$, $K_2(t;s)$, we adopt the convention:

$$[K_1, K_2] = \int_0^T \int_0^t \mathrm{Tr.}(GG^*)^{-1} K_1(t;s)(GG^*) K_2(t;s)^* \, ds \, dt \quad (8.22a)$$

With an obvious regrouping and using the fact that

$$K(\theta_o) = L(\theta_o)(I + K(\theta_o))$$

we can cast (8.22) into the form:

$$\begin{aligned} T \cdot q(\theta;T) = & \|(I - L(\theta))(m(\theta;.) - m(\theta_o;.))\|^2 \\ & + [L(\theta)(I + K(\theta_o)), (L(\theta) - L(\theta_o))(I + K(\theta_o))] \\ & - \| m(\theta_o;.)\|^2 - [K(\theta_o), L(\theta)(I+K(\theta_o))] \end{aligned} \quad (8.23)$$

from which it follows in particular that at $\theta=\theta_o$ we only have the last two terms, which in turn can be expressed:

$$T.q(\theta_o;T) = - \| m(\theta_o;.)\|^2 - [K(\theta_o), K(\theta_o)] \quad (8.24)$$

In a similar fashion, we can calculate, differentiating (8.21), (equival. (8.23) ! ) :

$$q_i(\theta;T) = \frac{2}{T}[(I-L(\theta))\, m_i(\theta;.),\ (I-L(\theta))(m(\theta;.) - m(\theta_o;.))]$$
$$+ \frac{2}{T}[L_i(\theta)(m(\theta_o;.) - m(\theta;.)),\ (I-L(\theta))(m(\theta;.) - m(\theta_o;.))]$$
$$+ \frac{2}{T}[L_i(\theta)(I+K(\theta_o)),\ (L(\theta) - L(\theta_o))(I+K(\theta_o)] \qquad (8.25)$$

where

$$m_i(\theta;t) = \frac{\partial}{\partial\theta_i} m(\theta;t)$$

$L_i(\theta)$ is the operator with kernel equal to the partial derivative with respect to $\theta_i$ of the kernel corresponding to $L(\theta)$

Note that $q_i(\theta_o;T)$ is zero.

Finally, we calculate:

$$q_{ji}(\theta;T) = \frac{2}{T}[(I-L(\theta))\, m_{ji}(\theta;.) + L_{ji}(\theta)(m(\theta_o;.) - m(\theta;.)) - L_j(\theta)m_i(\theta;.)$$
$$- L_i(\theta)\ m_j(\theta;.),\ (I-L(\theta))(m(\theta;.) - m(\theta_o;.))]$$
$$+ \frac{2}{T}[\ (I-L(\theta))\, m_i(\theta;.) + L_i(\theta)(m(\theta_o;.) - m(\theta;.)),\ (I-L(\theta))\, m_j(\theta;.)$$
$$+ L_j(\theta)(m(\theta_o;.) - m(\theta;.))] + \frac{2}{T}[L_i(\theta)(I+K(\theta_o)),\ L_j(\theta)(I+K(\theta_o))]$$
$$+ \frac{2}{T}[L_{ji}(\theta)(I+K(\theta_o))\ \ (L(\theta) - L(\theta_o))(I+K(\theta_o))] \qquad (8.26)$$

where $L_{ji}(\theta)$, $m_{ji}(\theta;.)$ denote the second partial derivatives.

In particular, at $\theta=\theta_o$,

$$q_{ji}(\theta_o; T) = \frac{2}{T}[L_i(\theta_o)(I+K(\theta_o)),\ L_j(\theta_o)(I+K(\theta_o))]$$
$$+ \frac{2}{T}[(I-L(\theta_o))\, m_i(\theta_o;.),\ (I-L(\theta_o))\, m_j(\theta_o;.)] \qquad (8.27)$$

Next, under the conditions (8.19) and (8.20), we can readily verify that $q(\theta; T)$, $q_i(\theta;T)$, $q_{ji}(\theta, T)$ all converge as T goes to infinity. Moreover, in taking the limits as T goes to infinity, we can replace the operator $L(\theta)$ by an operator with the 'stationary' kernel:

$$\tilde{L}(\theta)f = g; \qquad g(t) = \int_0^t \tilde{L}(\theta;t-s)\, f(s)\, ds$$

$$\tilde{L}(\theta;t) = C(\exp(A - PC^*(GG^*)^{-1}C)t)\, PC^*(GG^*)^{-1} \tag{8.28}$$

and similarly, $K(\theta)$ by the operator with the stationary kernel:

$$\tilde{K}(\theta)f = g; \qquad g(t) = \int_0^t \tilde{K}(\theta;t-s)\; f(s)\, ds$$

$$\tilde{K}(\theta;t) = C(\exp At) \quad PC^*(GG^*)^{-1} \tag{8.29}$$

Using the notation:

$$[f, g]_{av} \quad \text{for} \ \lim_{T\to\infty} \frac{1}{T}\int_0^T [(GG^*)^{-1}\; f(t),\; g(t)]\; dt$$

and for operators with stationary kernels [in $L_2\,(0, \infty)$]:

$$[K_1\; K_2]_\infty = \int_0^\infty \text{Tr.}\; (GG^*)^{-1}\; K_1(t)\; (GG^*)\; K_2(t)^* \; dt \tag{8.29a}$$

we can proceed to indicate

$$q(\theta) = \text{Lim}\;\; q(\theta;T)$$

$$q_i(\theta) = \text{Lim}\;\; q_i(\theta;T)$$

$$q_{ji}(\theta) = \text{Lim}\;\; q_{ji}(\theta;T)$$

It being understood that $\tilde{L}(\theta)$, $\tilde{K}(\theta)$ are now given by (8.28) and (8.29) respectively, we have:

$$q(\theta) = \|\, (I-\tilde{L}(\theta))(m(\theta;.) - m(\theta_o;.))\|^2_{av.} - \|\, m(\theta_o;.)\|^2_{av}$$

$$+[\tilde{L}(\theta)(I+\tilde{K}(\theta_o)),\ (\tilde{L}(\theta)-\tilde{L}(\theta_o))(I+\tilde{K}(\theta_o))]_\infty$$

$$-[\tilde{K}(\theta_o),\ \tilde{L}(\theta)(I+\tilde{K}(\theta_o))]_\infty \qquad (8.29b)$$

The others can be ovviously written down in similar manner. Of particular importance to us is the matrix with components:

$$q_{ji}(\theta_o) = 2[\tilde{L}_i(\theta_o)(I+\tilde{K}(\theta_o)),\ \tilde{L}_j(\theta_o)(I+\tilde{K}(\theta_o))]_\infty$$

$$+\ 2\,[(I-\tilde{L}(\theta_o))m_i(\theta_o;.),\ (I-\tilde{L}(\theta_o))\,m_j(\theta_o;.)]_{av} \qquad (8.30)$$

For, we can see, (as we should expect) that:

$$q_i(\theta) = \frac{\partial}{\partial\theta_i}\, q(\theta)$$

$$q_{ji}(\theta) = \frac{\partial^2}{\partial\theta_j\,\partial\theta_i}\, q(\theta)$$

and hence if the matrix in (8.30) is positive definite, then $q(\theta)$ is strictly convex in a non-zero neighborhood of $\theta_o$, and $\theta_o$ is the only root of the gradient $\nabla_\theta q(\theta)$ in that neighborhood. In fact, this is the motivation behind our "identification" technique. We have only to add to this that $q(\theta)$ can be approximated by $q(\theta;T)$ for large T, and what is more, by $q(\theta;y(.;\omega);T)$ itself, thanks to the "ergodicity" properties of this functional, as we shall prove next.

<u>Theorem 8.2</u> Assuming (8.19) and (8.20) :

$$E(\,|q(\theta;y(.;\omega);T) - q(\theta)|^2)$$

goes to zero as T goes to infinity, and moreover the convergence is uniform with respect to $\theta$ in compact subsets.

<u>Proof:</u> It is clearly enough to show that:

$$E(\,|\,q(\theta;y(.;\omega);T) - q(\theta;T)\,|^2) \qquad (8.31)$$

goes to zero uniformly in compact sets of $\theta$. Now

$$q(\theta;y(.;\omega);T) - q(\theta;T)$$

$$= (2/T)\ [m(\theta;.) + L(\theta)(m(\theta_o;.) - m(\theta;.)),\ L(\theta)(I+K(\theta_o))dZ(\theta_o;.;\omega)$$

$$- (I+K(\theta_o))dZ(\theta_o;.;\omega)]\ -(2/T)\ [m(\theta_o;.),\ L(\theta)(I+K(\theta_o))dZ(\theta_o;.;\omega)]$$

$$+(1/T)\ \Big\{[L(\theta)(I+K(\theta_o))dZ(\theta_o;.;\omega),\ L(\theta)(I+K(\theta_o))dZ(\theta_o;.;\omega)]$$

$$-[L(\theta)(I+K(\theta_o)),\ L(\theta)(I+K(\theta_o))]\Big\}$$

$$- (2/T)\ \Big\{[\ L(\theta)(I+K(\theta_o))dZ(\theta_o;.;\omega),\ (I+K(\theta_o))dZ(\theta_o;.;\omega)]$$

$$-[L(\theta)(I+K(\theta_o)),\ K(\theta_o)]\Big\}$$

This expression has three different kinds of integrals involving random processes. First, we have the kind:

$$\frac{1}{T}\int_0^T [k(\theta;t), dZ(\theta_o;t;\omega)]\ dt \quad ; \quad \lim_T \frac{1}{T}\int_0^T \|k(\theta;t)\|^2 dt < \infty$$

For this case, we have directly:

$$E(\ ((1/T)\int_0^T [\ k(\theta;t), dZ(\theta_o;t;\omega)]dt)^2.)$$

$$= (1/T)\ \Big\{(1/T)\ \int_0^T \|k(\theta;t)\|^2\ dt\Big\}$$

Since the "time average" in the curly brackets converges to a continuous function of $\theta$, this converges to zero uniformly for $\theta$ in compact sets.

Second, we have the form:

$$\frac{1}{T}\int_0^T [k(\theta;t), j(\theta;t;\omega)]\ dt$$

where

$$j(\theta;.;\omega) \sim L(\theta)(I+K(\theta_o))dZ(\theta_o;.;\omega)\ \text{ or }\ K(\theta_o)dZ(\theta_o;.;\omega)$$

Let

$$R(\theta;t;s) = E(\, j(\theta;t;\omega)\; j(\theta;s;\omega)^*)$$

Then

$$(\frac{1}{T^2})\, E(\, (\int_0^T [k(\theta;t), j(\theta;t;\omega)]\, dt)^2) \qquad \cdots\cdots\cdots \qquad (8.32)$$

$$= (1/T^2) \int_0^T \int_0^T [k(\theta;t), R(\theta;t;s)k(\theta;s)]\, ds\ dt$$

$$\leq \lambda(\theta;T)\,(1/T^2) \int_0^T \|k(\theta;t)\|^2\, dt$$

where $\lambda(\theta;T)$ denotes the largest eigenvalue of the operator:

$$R(\theta;T)f = g; \quad g(t) = \int_0^T R(\theta;t;s)\, f(s)ds \qquad 0 \leq t \leq T$$

and is a non-negative definite operator mapping $L_2(0, T)$ into itself. Since

$$\lim\, (1/T) \int_0^T \|k(\theta;t)\|^2 dt$$

exists and is finite, (8.32) will go to zero as required provided we can show that $\lambda(\theta;T)/T$ goes to zero, uniformly in $\theta$. But this follows from the fact that the process $j(\theta;t;\omega)$ is eventually stationary; that is to say:

$$\lim_T\ R(\theta;t+T;s+T) = R(\theta;t-s)$$

and further

$$R(\theta;t) = \int_{-\infty}^{\infty} e^{2\pi i f t}\ P(\theta;f)df$$

where $P(\theta;f)$ is bounded in $f$ and $\theta$. The argument in outline is this: for $T$ large enough, for any $k(.)$ in $L_2(0,T)$, of norm one:

$$(1/T)\,[R(\theta;T)k,k]$$

$$\sim (1/T)\int_0^T\int_0^T [R(\theta;t-s)\,k(s),\ k(t)]\,ds\,dt$$

$$= (1/T)\int_{-\infty}^{\infty} P(\theta;f)\int_0^T\int_0^T [e^{2\pi i f(t-s)}\,k(s),k(t)]\,ds\,dt\,df$$

$$\leq (1/T)\int_{-\infty}^{\infty} \|P(\theta;f)\|\ \left\|\int_0^T e^{2\pi i f t}\,k(t)\,dt\right\|^2 df$$

$$\leq (1/T)\ \sup_{\theta;f}\ \|P(\theta;f)\|$$

$$= (1/T)\ \sup_{\theta;f}\ \|P(\theta;f)\|$$

Or,

$$\lambda(\theta;T)/T \leq (1/T)\ \sup_{\theta;f}\ \|P(\theta;f)\| \longrightarrow 0$$

uniformly in $\theta$.

Next, we have the form

$$(1/T)\int_0^T [j(\theta;s;\omega)\ ,\ dZ(\theta_o;s;\omega)]$$

Here it is immediate that:

$$E\left(\left((1/T)\int_0^T [j(\theta;s;\omega),\ dZ(\theta_o;s;\omega)]\right)^2\right)$$

$$= \left(\frac{1}{T^2}\right)\int_0^T E(\|j(\theta;s;\omega)\|^2)\ ds$$

$$\longrightarrow 0 \text{ uniformly in } \theta \text{ in compact sets.}$$

Finally, we need to consider the form:

$$(1/T)\int_0^T [k_i(\theta;s;\omega),\ k_j(\theta;s;\omega)]\ ds - (1/T)\int_0^T E([k_i(\theta;s;\omega), k_j(\theta;s;\omega)])\ ds,$$

$$i = 1, 2\ ; \quad j = 1, 2$$

where

$$k_1(\theta;.;\omega) \sim L(\theta)[I+K(\theta_o)]\ dZ(\theta_o;.;\omega)$$

$$k_2(\theta;.;\omega) = h(\theta_o;.;\omega)$$

But the square of the expected value of this expression:

$$= (1/T)^2 \int_0^T\int_0^T \Big\{ E([\ k_i(\theta;s;\omega),\ k_j(\theta;s;\omega)]\ [k_i(\theta;t;\omega),\ k_j(\theta;t;\omega)])$$

$$- E([k_i(\theta;s;\omega),\ k_j(\theta;s;\omega)])\ E([\ k_i(\theta;t;\omega), k_j(\theta;t;\omega)]) \Big\}\ ds\ dt \qquad (8.33)$$

Consider first the case where $k_i(..)$, $k_j(...)$ are one-dimensional. Then, by the rules for calculating "four" products of Gaussians, we have:

$$E(k_i(\theta;s;\omega)\ k_j(\theta;s;\omega)\ k_i(\theta;t;\omega)k_j(\theta;t;\omega))$$

$$- E(k_i(\theta;s;\omega)\ k_j(\theta;s;\omega))\ \ E(k_i(\theta;t;\omega)\ k_j(\theta;t;\omega))$$

$$= E(k_i(\theta;s;\omega)\ k_j(\theta;t;\omega))\ .\ E(k_j(\theta;s;\omega)\ k_i(\theta;t;\omega))$$

$$+ E(\ k_i(\theta;s;\omega)\ k_i(\theta;t;\omega)\ )\ E(k_j(\theta;s;\omega)\ k_j(\theta;t;\omega))$$

In the vector case we have a finite sum of such expressions. Let us next note that the processes $k_1(\theta;t;\omega)$ and $k_2(\theta;t;\omega)$ are eventually 'stationary' and 'stationarily related'; that is to say, letting

$$R_{ij}(\theta;t;s) = E(k_i(\theta;t;\omega)\ k_j(\theta;s;\omega)^*)$$

we have:

$$\lim_{T} R_{ij}(\theta;t+T;s+T) = R_{ij}(\theta;t-s)$$

uniformly for $\theta$ in compact sets. Because of this, considering first again the one-dimensional case, we have:

$$\lim_{T} (1/T) \int_0^T \int_0^T R_{ij}(\theta;t;s) R_{ij}(\theta;s;t) \, ds \, dt$$

$$= \lim_{T} (1/T) \int_0^T \int_0^T R_{ij}(\theta;t-s) R_{ij}(\theta;s-t) \, ds \, dt$$

and the second limit is readily calculated from the kernels $\tilde{L}(\theta;t)$ and $\tilde{K}(\theta;t)$, and hence

$$\lim_{T} (1/T^2) \int_0^T \int_0^T R_{ij}(\theta;t;s) R_{ij}(\theta;s;t) \, ds \, dt$$

goes to zero uniformly in $\theta$ in compact sets. The vector case follows since it can be expressed as a finite sum of such expressions, and thus (8.33) goes to zero uniformly in $\theta$ as required.

Finally, since only the same kinds of integrals as herein are involved, we can state without further proof:

Corollary: An analagous statement to Theorem 8.2 holds for all derivatives of $q(\theta;y(\cdot;\omega);T)$.

The condition that the matrix defined by (8.30) is positive definite plays a crucial role in what follows. We shall denote this condition S'. First we note:

<u>Theorem 8.3</u> The condition S' implies condition S

Proof   Suppose $\theta_n$ is a sequence converging to $\theta_o$ such that the responses are the same. That is to say, we have:

$$y(t;\omega) = C_n \int_0^t x(s;\omega)\, ds + D_n \int_0^t u(s)ds + G\, W(t;\omega)$$

$$x(t;\omega) = A_n \int_0^t x(s;\omega)ds + B_n \int_0^t u(s)ds + F_n\, W(t;\omega)$$

where the subscript n indicates $\theta_n$. Taking expected values, we obtain that

$$m(\theta_n;t) = m(\theta_o;t)$$

Next

$$dy(t;\omega) = m(\theta_n;t)dt + (I+K(\theta_n))\, dZ(\theta_n;t;\omega)$$

where $Z(\theta_n;t;\omega)$ is a Wiener process with covariance matrix (GG*). Hence

$$(I+K(\theta_n))\, dZ(\theta_n;\cdot;\ ) \quad = \quad d\tilde{y}(\theta_o;\cdot;\omega) = (I+K(\theta_o))\, dZ(\theta_o;\cdot;\omega)$$

or

$$(I+K(\theta_o))^{-1}\, (I+K(\theta_n))\, dZ(\theta_n;\cdot;\omega) = dZ(\theta_o;\cdot;\omega)$$

Observe now that the two Wiener processes $Z(\theta_n;\cdot;\omega)$ and $Z(\theta_o;\cdot;\omega)$ are both measurable with respect to the same growing sigma algebra, namely $\beta_y(t)$, and moreover the sigma-algebra generated by each process is equivalent to $\beta_y(t)$. Hence, as in Chapter VI, it follows that ((6.2) being satisfied):

$$dZ(\theta_n;t;\omega) = M(t)\, dZ(\theta_o;t;\omega)$$

and $M(t)$ must be non-singular and:

$$M(t)\, M(t)^* = \text{Identity matrix}$$

Hence the operator

$$(I + K(\theta_o))^{-1}\,(I+K(\theta_n))$$

corresponds to multiplication by the matrix $M(t)$; and since $K(\theta_o)$, $K(\theta_n)$ are Volterra operators of Hilbert-Schmidt type, it follows that $M(t)$ must be the identity matrix, and hence that

$$K(\theta_o) = K(\theta_n)$$

and hence also

$$L(\theta_o) = L(\theta_n)$$

From (8.24) it then follows that

$$q(\theta_n;T) = q(\theta_o;T)$$

Certainly for large enough $n$, the condition (8.19) is satisfied, and hence taking limits in $T$, we get

$$q(\theta_n) = q(\theta_o)$$

for all $n$ sufficiently large. But this is a contradiction, since if condition S' is satisfied we know that there is a non-zero neighborhood of $\theta_o$ in which $q(\theta)$ is strictly convex and $\theta_o$ is the only point of minimum therein.

We now come to our main result which asserts that the root of the gradient of $q(\theta;y(\cdot;\omega);T)$ converges in probability to $\theta_o$. It is patterned on the analagous classical result as given by Cramer [14].

Theorem 8.4 Suppose condition S' holds, as well as (8.19), (8.20). Then given any arbitrarily small positive quantities $\delta, \epsilon$, the gradient $\nabla_\theta q(\theta;y(\cdot;\omega);T)$ has a root in a sphere of radius $\delta$ about $\theta_o$, with probability exceeding $(1-\epsilon)$, for all $T > T(\delta;\epsilon)$.

Proof We need a 'Taylor series with remainder' for $\nabla_\theta q(\theta;y(\cdot;\omega);T)$ for each T. For this, let $Q(\theta;y(\cdot;\omega);T)$ denote the matrix with components

$$q_{ij}(\theta;y(\cdot;\omega);T)$$

Let $J(\theta;y(\cdot;\omega);T)$ denote the gradient (Frechet derivative) of $Q(\theta;y(\cdot;\omega);T)$ with respect to $\theta$; $J(\theta;y(\cdot;\omega);T)$ is thus a linear transformation of the parameter $\theta$ space into the space of square matrices. Then we have:

$$\nabla_\theta q(\theta;y(\cdot;\omega);T) - \nabla_\theta q(\theta_o;y(\cdot;\omega);T) = \int_0^1 \frac{d}{ds}\nabla_\theta q(((1-s)\theta_o+s\theta);y(\cdot;\omega);T)ds$$

$$= \int_0^1 Q(((1-s)\theta_o+s\theta);y(\cdot;\omega);T)ds\ (\theta - \theta_o)$$

And in a similar fashion:

$$Q(((1-s)\theta_o + s\theta);y(\cdot;\omega);T)$$

$$= Q(\theta_o;y(\cdot;\omega);T) + \int_0^1 J(((1-st)\theta_o+st\ \theta);y(\cdot;\omega);T)dt\ (1-s)(\theta - \theta_o)$$

Let

$$\overline{J}(\theta;y(\cdot;\omega);T) = \int_0^1\int_0^1 J(((1-ts)\theta_o+st\ \theta);y(\cdot;\omega);T)(1-s)ds\ dt$$

Then we can finally write:

$$\nabla_\theta q(\theta;y(\cdot;\omega);T) = \nabla_\theta q(\theta_o;y(\cdot;\omega);T) + Q(\theta_o;y(\cdot;\omega);T)(\theta-\theta_o)$$

$$+ \left( \bar{J}(\theta;y(\cdot;\omega);T)(\theta-\theta_o)\right)(\theta-\theta_o) \tag{8.34}$$

The basic idea can be heuristically explained in terms of the one-dimensional version of (8.34). Let us fix the $\omega$, so that we can write the right side as

$$ax^2 + bx + c \; ; \; x = \theta - \theta_o$$

where we note that for all $T$ large enough, $b \geq m > 0$, and $c$ goes to zero with $T$, and $|a|$ is bounded for all $T$. The case where $|a|$ goes to zero being quite simple, we shall only need to discuss the case where $|a|$ is bounded away from zero. Then we know that the roots are given by:

$$(-b \pm \sqrt{b^2 - 4ac}\,)/2a$$

and since $|ac|/b^2$ can be made as small as we wish compared to unity, we can approximate the roots as:

$$\approx ((-b \pm b(1 - 2ac/b^2))/2a$$

and hence one of them as:

$$\approx (-b+(b - \frac{2ac}{b^2}))/2a = c/b$$

and since $|c|$ can be made small independent of $T$, the root can be made to lie in the interval $[-\delta, \delta]$.

To handle the general case we now proceed formally as follows. We use the convergence in the mean square (and hence in probability) and as is common in numerical analysis, a simple version of a fixed point theorem to show existence of a root.

Let $\epsilon$, $\delta_1, \delta_2$ be arbitrary positive numbers. By Theorem 8.2 and Corollary, the coefficients in (8.34) all converge in the mean square sense, uniformly in $\theta$ in the compact set $\bar{\mathcal{N}}$. (Henceforth we shall simply say 'uniformly in $\theta$' to mean this.) Hence for every $\omega$ in a set $\Lambda(\epsilon)$ of measure exceeding $(1-\epsilon)$, and for every $\theta$ in $\bar{\mathcal{N}}$.

$$\|\nabla_\theta q(\theta_o;y(\cdot;\omega);T) - \nabla_\theta q(\theta_o)\| = \|\nabla_\theta q(\theta_o;y(\cdot;\omega);T)\| < \delta_1;$$

$$\|Q(\theta_o;y(\cdot;\omega);T) - Q(\theta_o)\| < \delta_2$$

$$\|\bar{J}(\theta;y(\cdot;\omega);T) - \bar{J}(\theta)\| < \delta_2$$

for all $T > T(\epsilon;\delta_1;\delta_2)$. From now on, we shall only consider $\omega$ in $\Lambda(\epsilon)$ and $T > T(\epsilon;\delta_1;\delta_2)$. Hence for $\delta_2$ sufficiently small, so is $Q(\theta_o;y(\cdot;\omega);T)$ for all $\omega$ in $\Lambda(\epsilon)$, $T > T(\epsilon;\delta_1,\delta_2)$.
More precisely, the smallest eigenvalue of the matrix above is bigger than some positive number m say. Hence the inverse

$$(Q(\theta_o;y(\cdot;\omega);T))^{-1}$$

has its largest eigen-value less than $(1/m)$. Hence

$$\|(Q(\theta_o;y(\cdot;\omega);T)^{-1} \nabla_\theta q(\theta_o;y(\cdot;\omega);T)\| < \delta_1/m \ldots \qquad (8.35)$$

Also:

$$\|Q(\theta_o;y(\cdot;\omega);T)^{-1} \quad \overline{J}(\theta \ ;y(\cdot;\omega);T)(\theta-\theta_o)\| < (M/m)\|\theta-\theta_o\| \tag{8.36}$$

since because the convergence in $\theta$ is uniform, and $\overline{J}(\theta)$ is bounded, we can take

$$\|\overline{J}(\theta;y(\cdot;\omega);T)\| < M < \infty \tag{8.37}$$

Next, to apply the fixed point theorem, let

$$x = \theta - \theta_o$$

Define:

$$f(x) = x - (Q(\theta_o;y(\cdot;\omega);T))^{-1} \quad \nabla_\theta q(x+\theta_o;y(\cdot;\omega);T) \tag{8.38}$$

Substituting this into (8.34), we have

$$f(x) = Q(\theta_o;y(\cdot;\omega);T)^{-1} \quad \nabla_\theta q(\theta_o;y(\cdot;\omega);T)$$

$$+ Q(\theta_o;y(\cdot;\omega);T)^{-1} \ (\overline{J}(\theta;y(\cdot;\omega);T)(x))(x)$$

Using our estimates (8.35), (8.36), (8.37), we get

$$\|f(x)\| \leq \delta_1/m + (M/m)\|x\|^2$$

Choose $\delta$ so that

$$\delta < (m/2M)$$

and then choose

$$\delta_1 = m\,\delta/4$$

Then for all $\|x\| < \delta$,

$$\|f(x)\| \leq \delta/4 + \delta(1/2) = \frac{3\delta}{4}$$

and hence the fixed point theorem applies. [Of course we are assuming also that $\delta$ is small enough so that the sphere of radius less than $\delta$ lies in $\mathcal{N}$.] Hence there exists a point $x_r$ say, in this sphere so that

$$f(x_r) = x_r$$

Hence from (8.38) we see that

$$\nabla_\theta q(\theta_r; y(\cdot;\omega); T) = 0$$

Since $\delta_2$ can be fixed, we see that we have the required root for all $T > T(\varepsilon;\delta)$ in a sphere of radius $\delta$ about $\theta_o$ as we needed to show.

Remark A computational algorithm for finding $\theta_T$ based more or less on the Theorem is:

$$\theta_{n+1} = \theta_n - R(\theta_n; y(\cdot;\omega); T)^{-1} \nabla_\theta q(\theta_n; y(\cdot;\omega); T) \tag{8.40}$$

where $R(\theta_n; y(\cdot;\omega); T)$ is the matrix with components:

$$(2/T)[(I-L(\theta_n))m_i(\theta_n;\cdot) + L_i(\theta_n)(dy(\cdot;\omega)-m(\theta_n;\cdot)),$$

$$(I-L(\theta_n))\, m_j(\theta_n;\cdot) + L_j(\theta_n)(dy(\cdot;\omega)-m(\theta_n;\cdot))] \tag{8.41}$$

We can assert that under the conditions of the Theorem, for $\omega$ in $\Lambda(\varepsilon)$, there is a neighborhood of $\theta_o$ such that if we start with $\theta_1$ in that neighborhood, then for all $T > T(\varepsilon;\delta)$, $\theta_n$ will converge to a root $\theta_T$.

Remark: In calculating the gradient $\nabla_\theta q(\theta;y(.;\,);T)$, we may, in (8.40) clearly use the stationary forms: use P in place of P(t) so that:

$$q_i(\theta;y(.;\,);T) \doteq (2/T)\left[(I-\tilde{L}(\theta))\, m_i(\theta;.) + \tilde{L}_i(\theta)(dy(.;\omega) - m(\theta;.)),\ (I-\tilde{L}(\theta))(m(\theta;.)-dy(.;\omega))\right]$$

We observe next that the matrix $Q(\theta_o)$ is the sum of two non-negative definite matrices. Actually we can state:

Theorem 8.5 Let M denote the matrix with components

$$[m_i(\theta_o;\cdot),\ m_j(\theta_o;\cdot)]_{av.}$$

Let $\mathscr{L}$ denote the matrix with components:

$$[L_i^{\sim}(\theta_o),\ L_j^{\sim}(\theta_o)]_{\infty}$$

Then the matrix $Q(\theta_o)$ is non-singular if and only if either M or $\mathscr{L}$ is non-singular.

Proof Suppose M is singular. Then

$$\|\Sigma\, a_i\, m_i(\theta_o;\cdot)\|^2_{av.} = 0$$

for some $\{a_i\}$ not all zero. Then clearly the same holds for

$$[\Sigma\, a_i(I-\tilde{L}(\theta_o))m_i(\theta_o;\cdot),\ \Sigma a_j(I-\tilde{L}(\theta_o))m_j(\theta_o;\cdot)]_{av.}$$

and hence

$$\left[(I-\tilde{L}(\theta_o))m_i(\theta_o;.),\ (I-\tilde{L}(\theta_o))m_j(\theta_o;.)\right]_{av.} \tag{8.42}$$

is singular also.

Conversely, suppose (8.42) is singular. Then

$$\Sigma a_i (I-\tilde{L}(\theta_o)) m_i(\theta_o;\cdot) = g(\cdot)$$

for constants $\{a_i\}$ not all zero is such that

$$\| g(\cdot) \|^2_{av.} = 0$$

But from

$$(I + K(\theta_o))^{-1} = (I - L(\theta_o))$$

for every T, it also follows that

$$(I + \tilde{K}(\theta_o))^{-1} = (I - \tilde{L}(\theta_o))$$

Hence

$$(I + \tilde{K}(\theta_o))\, g(\cdot) = \Sigma\, a_i\, m_i(\theta_o;\cdot)$$

But

$$\| (I + \tilde{K}(\theta_o))\, g(\cdot) \|^2_{av.} = 0$$

Hence M is singular also.

Similarly we can prove that $\mathscr{L}$ is singular if and only if

$$[\tilde{L}_i(\theta_o)(I + \tilde{K}(\theta_o)),\ \tilde{L}_j(\theta_o)(I + \tilde{K}(\theta_o))]_\infty \qquad (8.43)$$

is singular. But $Q(\theta_o)$ is the sum of the non-negative definite matrices (8.41) and (8.42) and is non-singular if and only if at least one of them is. This proves the Theorem.

Of special interest is the case where the matrix M is non-singular. This can happen for instance if F contains no unknown parameters. We can then simplify the algorithm (8.40) considerably.

Let

$$f_i(\theta;y(\cdot;\omega);T) = \frac{1}{T}[(I - L(\theta))m_i(\theta;\cdot), (I - L(\theta))(m(\theta;\cdot)-dy(\cdot;\omega))] \qquad (8.43)$$

Then we can verify by the same technique as before that

$$f_i(\theta) = \lim_T f_i(\theta;y(\cdot;\omega);T) = [(I - \tilde{L}(\theta))(m(\theta;\cdot) - m(\theta_o;\cdot)),$$

$$(I - \tilde{L}(\theta))m_i(\theta;\cdot)]_{av}.$$

Proceeding in a similar manner, we can also calculate that:

$$\lim \frac{\partial}{\partial\theta_j} f_i(\theta;y(\cdot;\omega);T)$$

$$= [(I - \tilde{L}(\theta))\, m_i(\theta;\cdot),\ (I - \tilde{L}(\theta))\, m_j(\theta;\cdot)]_{av}.$$

as in Theorem 8.4.

Remark It should be noted that the calculation of the partial derivative (8.43) is considerably simpler than (8.25).

Remark As before, a computational algorithm based on this Theorem is:

$$\theta_{n+1} = \theta_n - M(\theta_n;T)^{-1}\, F(\theta_n;y(\cdot;\omega);T)$$

where $M(\theta_n;T)$ is the matrix with components:

$$(1/T)[(I - L(\theta_n))\, m_i(\theta_n;\cdot), (I - L(\theta_n))\, m_j(\theta_n;\cdot)]$$

We can make the same assertion about this algorithm as before. The algorithm is of course much simpler than before. An even greater possible simplification would be to take, instead of $M(\theta_n;T)$, the matrix with components:

$$(1/T)\,[m_i(\theta_n;\cdot),\ m_j(\theta_n;\cdot)]\,.$$

And similarly the matrix

$$(2/T)\left[\tilde{L}_i(\theta_n),\ \tilde{L}_j(\theta_n)\right] \quad + \quad (2/T)\left[m_i(\theta_n;.),\ m_j(\theta_n;.)\right]$$

instead of the matrix $R(\theta_n;y(.;\omega);T)$ (in (8.40)).

$$+\,[(I - \tilde{L}(\theta))\,\frac{\partial^2}{\partial\theta_j\partial\theta_i}\ m(\theta;\cdot),\ (I - \tilde{L}(\theta))(m(\theta;\cdot) - m(\theta_o;\cdot))]_{av.}$$

$$-\,[(I - \tilde{L}(\theta))\ m_i(\theta;\cdot),\ \tilde{L}_j(\theta))(m(\theta;.)-m(\theta_o;\cdot))]_{av.}$$

$$-\,[(\tilde{L}_j(\theta))\ (m_i(\theta;\cdot)),(I - \tilde{L}(\theta))(m(\theta;\cdot)-m(\theta_o;\cdot))]_{av.}$$

Then we can prove a result analagous to Theorem 8.4.

<u>Theorem 8.6</u> Suppose the matrix $M$ is positive definite. Then denoting by $F(\theta;y(\cdot;\omega);T)$ the vector with components

$$f_i(\theta;y(\cdot;\omega);T)$$

we have that: given $\epsilon > 0$, $\delta > 0$, then with probability exceeding $(1-\epsilon)$, $F(\theta;y(.;\omega);T)$ has a root in a sphere of radius $\delta$ about $\theta_o$ for all $T > T(\epsilon;\delta)$.

Proof We note that $M$ being positive definite implies that the matrix with components

$$\lim \frac{\partial}{\partial \theta_j} f_i(\theta; y(\cdot;\omega); T)$$

is also positive definite at $\theta = \theta_o$. We can therefore clearly proceed as in Theorem 8.4.

## APPENDIX I

In this Appendix we collect together some properties of Volterra operators used explicitly or implicitly in the text. Details of proofs are given only when not readily available in the literature.

Let $H = L_2([0,1]; E_n)$ denote the (real Hilbert space) $L_2$ - space of n-by-one real valued functions on the interval $[0,1]$. By a Volterra operator we mean (for our purposes) an integral operator mapping H into H of the form:

$$L f = g; \quad g(t) = \int_0^t L(t;s) f(s) \, ds$$

where $L(t;s)$ is continuous in the triangle $0 \leq s \leq t \leq 1$. Thus defined, L is clearly Hilbert-Schmidt, and has the characteristic property of Volterra operators that: the Neumann expansion

$$(\lambda I - L)^{-1} = \sum_0^\infty L^n / \lambda^{n+1}$$

is valid for all non-zero $\lambda$.

Of importance to us is the question: when is L of trace-class? Let us recall that an operator L is trace-class if

$$\sum_1^\infty [R\phi_n, \phi_n] < \infty$$

where R is the positive square root of $L^*L$ and $\{\phi_n\}$ is the complete orthonormal sequence of eigen-vectors of R (including those corresponding to zero eigenvalues). If L is trace-class, then

$$\sum_1^\infty |[L g_n, g_n]| < \infty$$

for any orthonormal sequence $\{g_n\}$, and we define

$$\text{Trace } L = \sum_1^\infty [Lg_n, g_n]$$

where $\{g_n\}$ is any <u>complete</u> orthonormal system (and the sum is independent of the particular sequence chosen). First let us look at some useful necessary conditions.

Theorem: In order that L be trace-class it is necessary that

$$L(t;t) = L(t;t)^*; \ \text{Tr. } L(t;t) = 0$$

<u>Proof:</u> First of all let us note that if L is trace-class so is $L^*$. Moreover it is known (see [32]) that if a Volterra operator is trace-class, its trace must be zero. Since we are in a real Hilbert space,

$$\text{Tr. } L = \text{Tr. } L^*$$

and hence $(L+L^*)$ is also trace-class with zero trace. Let

$$K(t;s) = \begin{cases} L(t;s) & 0 < s < t \leq 1 \\ L(s;t)^* & 0 < t < s \leq 1 \end{cases}$$

Then we have:

$$(L+L^*)f = g; \ g(t) = \int_0^1 K(t;s)\, f(s)ds$$

Let $\{\phi_i\}$ denote the orthonormalized sequence of eigen-functions of the compact self-adjoint operator $(L+L^*)$. Since for any f in $\mathcal{H}$, we can readily verify that both Lf and $L^*f$ are continuous functions, it follows that the $\{\phi_i\}$ are continuous. Moreover, denoting the corresponding eigen-values by $\{\lambda_i\}$, we know that

$$L_b f = g; \quad g(t) = \int_0^t L(t;s)f(s)ds \quad 0 \leq t \leq b$$

defines $L_b$ as a Volterra operator mapping $H_b$ into itself. Moreover it is trace-class, since any orthonormal basis in $H_b$ provides us with an

$$K(t, s) = \sum_i \lambda_i \, \phi_i(t) \, \phi_i(s)^* \quad \text{a.e.} \quad \text{in} \quad 0 \leq s, t \leq 1.$$

Moreover, since $(L+L^*)$ is trace-class, we have that

$$\sum_1^\infty |\lambda_i| < \infty$$

Further since $L(t,s)$ is continuous $0 \leq s \leq t \leq 1$, we can readily estimate that

$$\sup_i \; \sup_t \; \|\phi_i(t)\| \leq c < \infty$$

Hence the series

$$\Sigma \lambda_i \, \phi_i(t) \, \phi_i(s)^*$$

converges uniformly in $0 \leq s, t \leq 1$. Hence

$$r(t;s) = \Sigma_i \, \phi_i(t) \, \phi_i(s)^*$$

is also continuous in $0 \leq s, t \leq 1$. But

$$r(t;s) = L(t;s) \quad s < t$$

$$= L(s;t)^* \quad t < s$$

Hence

$$r(t;t) = L(t;t) = L(t;t)^*$$

$$\int_0^1 \text{Tr. } L(t;t)\, dt = 0$$

Next let us observe that under the continuity conditions imposed on $L(t;s)$, taking any $b, 0 < b < 1$, and $H_b = L_2(0, b);E_n)$:

orthonormal system in H (by defining the functions to be zero for $b<t<1$). Hence we can apply the results so far obtained to deduce:

$$\int_0^b \text{Tr. } L(t;t)dt = 0$$

and since b is arbitrary that:

$$\text{Tr. } L(t;t) = 0$$

Remark: We note that under the continuity condition imposed on $L(t;s)$, we can state that if $(L+L^*)$ is trace-class, then

$$L(t;t) = L(t;t)^*$$

and that

$$\text{Tr. } (L+L^*) = \int_0^1 \text{Tr. } L(t;t)dt \ .$$

Here is an example of an operator which is not trace-class:

$$Lf = g; \ g(t) = \int_0^t f(s)ds$$

For,

$$(L+L^*) \, f \sim \int_0^1 f(s)ds$$

and thus $(L+L^*)$ has a finite dimensional range, and hence is trace-class (since $(L+L^*)^2$ then has a finite dimensional range). But

$$\text{Tr. } (L+L^*) = n \neq 0$$

Hence L cannot be trace-class. Note that for any operator L, if $(L+L^*)$ is trace-class, then for every complete orthonormal sequence $\{g_n\}$, we have

$$\sum_1^\infty [\, Lg_n \, , \, g_n] \quad = (1/2) \ \text{Tr. } (L+L^*)$$

where the series converges absolutely (and conversely, since $(L+L^*)$ is self-adjoint). But L need not be trace-class as the above example shows.

Finally, defining the operator

$$Lf = g \; ; \quad g(t) = A \int_0^t f(s)\, ds$$

we note that for $(L+L^*)$ to be trace class, it is necessary that $A$ be self-adjoint. And if $A$ is self-adjoint, $(L+L^*)$ has a finite dimensional range and hence is trace-class. But L itself is never trace-class. For, denoting the operator

$$\int_0^t f(s)\, ds$$

by $\mathcal{J}$ we note that

$$\sqrt{L^* L} = \sqrt{A^* A}\sqrt{\mathcal{J}^* \mathcal{J}}$$

and if $\phi_i$ denotes the orthonormalized eigen vectors of $\mathcal{J}^* \mathcal{J}$ with corresponding eigen-values $\lambda_i$, we have that unless $A$ is zero:

$$\sum_1^\infty \lambda_i [A^* A \phi_i , \phi_i] = \infty ,$$

since the $\phi_i$ can be taken in the form

$$\phi_i (t) = h_k(t) e_j$$

where $e_j$ are orthonormal unit vectors in $E_n$, so that

$$\sum_i \lambda_i [\sqrt{A^* A} \phi_i, \phi_i] = (\sum_j [\sqrt{A^* A} e_j, e_j]) (\Sigma \lambda_i)$$

$$= \text{Tr.} \sqrt{A^* A} \; \Sigma \lambda_i = + \infty$$

Next, let us look at sufficient conditions. Unfortunately, the necessary conditions are far from sufficient, as the following counter example (slightly modified from [15]) shows: According to T. Carleman's classical construction (see[15]) we can find a continuous periodic function $q(t)$, $0 \leq t \leq 1$, such that $q(t)$ has the Fourier expansion:

$$q(t) \sim 2 \sum_{1}^{\infty} c_k \cos 2\pi kt + c_o$$

where

$$\sum_{1}^{\infty} |c_k| = +\infty$$

With $E_n = E_1$, define the operator L by:

$$L f = g; \qquad g(t) = \int_o^t (q(t-s) - q(0)) f(s)\, ds$$

Then

$$(L+L^*)f \sim \int_o^1 (q(t-s) - q(0))\, f(s)\, ds$$

and hence, with

$$g_k(t) = \sqrt{2} \cos 2\pi kt$$

we have

$$[(L+L^*)\, g_k\, ,\, g_k] = c_k$$

so that $(L+L^*)$ cannot be trace-class. On the other hand, the corresponding function $L(t;t)$ is now identically zero. We state two sufficient conditions of interest to us.

<u>Theorem</u>: Suppose $L(t;s)$ has the form:

$$L(t;s) = \int_s^t M(t;\sigma)\, d\sigma$$

where

$$\int_0^1 \int_0^t ||M(t;\sigma)||^2 \, d\sigma \, dt < \infty$$

Then L is trace-class.

Proof: Define the operator P by:

$$P f = g \; ; \; g(t) = \int_0^t M(t;s) \, f(s) \, ds$$

Define the operator $\mathcal{J}$ by:

$$\mathcal{J} f = g \; ; \qquad g(t) = \int_0^t f(s) \, ds$$

Then we can verify by direct calculation that

$$L = P \mathcal{J}$$

and P, $\mathcal{J}$ are both Hilbert-Schmidt, and hence L is trace-class, since the product of Hilbert-Schmidt operators is trace-class.

Theorem: Suppose L(t;s) has the form

$$L(t;s) = \Phi(t) \, \Phi(s)^{-1} \, P(s)$$

where $\Phi(t)$ is a fundamental matrix solution of:

$$\dot{\Phi}(t) = A(t) \, \Phi(t)$$

where the matrix A(t) is continuous in t, say. Suppose P(s) is self-adjoint and continuously differentiable in s. Then (L+L*) is trace-class.

Proof Denote the operator Q by:

$$Qf = g; \; g(t) = \int_0^t A(s) \, f(s) \, ds$$

and the operator P by:

$$Pf = g; \; g(t) = \int_0^t P(s) \, f(s) \, ds$$

Then we can directly verify that

$$L = (I-Q)^{-1} P = P + \sum_{1}^{\infty} Q^n P$$

But $Q, P$ being Hilbert-Schmidt and Volterra, $Q^n P$ is trace-class for $n \geq 1$. Hence it is enough to show that $(P + P^*)$ is trace-class. But this follows from the fact that:

$$(P+P^*)f=g; \; g(t) = P(t) \int_0^1 f(s)ds - \int_0^t (P(t) - P(s))f(s)ds$$

and the second term, can be expressed:

$$P(t) - P(s) = \int_s^t \dot{P}(\sigma) \, d\sigma$$

so that the previous Theorem applies. Also we have:

$$\text{Tr } (L+L^*) = \int_0^1 \text{Tr. } P(s)ds$$

Theorem Let $t_k$ be such that $0 \leq t_k \leq 1$, $k = 0, 1, ..n$

$$t_i < t_{i+1} \; ; \; t_0 = 0 \; ; \; t_n = 1$$

Define the operator $L_n$ by:

$$L_n f = g \; ; \; g(t) = \int_o^{t_i} L(t_i;s) \, f(s) \, ds \; ; \; t_i \leq t < t_{i+1} \; ; \; i = 0, ..n-1$$

Then $L_n$ is trace-class, and the trace is zero.

Proof: Since the subdivision points are fixed, and $g(t)$ is a step function of the form:

$$g(t) = \text{constant vector}, \quad t_i \leq t < t_{i+1}, \quad i = 0, ..n-1,$$

it follows that the range of $L_n$ is finite dimensional, and hence trace-class. Let $\{g_n\}$ be a complete orthonormal sequence. Then

$$\sum_k [L_n g_k, g_k] = \sum_{i=0}^{n-1} \sum_k [\int_0^{t_i} L(t_i; s)\ g_k(s)\ ds, \int_{t_i}^{t_{i+1}} g_k(s)\ ds]$$

Now consider for each i,

$$\sum_k [\int_0^{t_i} L(t_i\ ; s)\ g_k(s)\ ds, \int_{t_i}^{t_{i+1}} g_k(s)\ ds]$$

First take the one-dimensional case. Define

$$h(t) = L(t_i; s) \quad 0 < t < t_i$$

$$= 0 \ \text{otherwise}$$

$$b(t) = 1\ , \ t_i < t < t_{i+1}$$

$$= 0 \ \text{otherwise}$$

Then

$\int_0^{t_i} L(t_i;s)\ g_k(s)\ ds$ being the Fourier coefficients of $h(.)$

$\int_{t_i}^{t_{i+1}} g_k(s)\ ds$ being the Fourier coefficients of $b(.)$

it follows that the sum above

$$= \int_0^1 h(t)\ b(t)\ dt = 0$$

thus verifying the answer in the one-dimensional case. The n-dimensional case can be handled by considering an orthonormal basis of the form

$$g_k(t) = h_i(t)\ e_j$$

where $h_i(.)$ is a complete orthonormal sequence in one dimension, and $e_j$ are orthonormal basis for $E_n$.

Finally, we prove the result needed in Chapter V.

<u>Theorem:</u> Let L be a volterra operator of Hilbert-Schmidt type. Suppose $(L + L^*)$ is trace-class. Then

$$\text{Log Det}\,[(I-L)(I-L^*)] = \text{Tr. Log}\,[(I-L)(I-L^*)] = \text{Tr. }(-1)(L+L^*) \qquad \text{(AI}$$

<u>Proof</u> We specialize a result given by Krein (15, p. 174) involving the perturbation determinant (Fredholm) for two bounded operators A, B of Hilbert-Schmidt type such that (A-B) is trace class. (Perhaps it should be remarked that we are not assuming that L is trace class. Of course if a Volterra operator is trace-class, its trace must be zero.):

$$\frac{d}{ds}\,\text{Log.Det}\,[(I-sB)(I-sA)^{-1}] = \text{Tr. }(A(s) - B(s)) \qquad \text{(AI}$$

where

$$I + sA(s) = (I - sA)^{-1}$$

$$I + sB(s) = (I - sB)^{-1}$$

and it is assumed that the indicated inverses exist and are bounded. Let us briefly indicate the proof of this, following Krein [15]. We have:

$$A(s) - B(s) = (I-sB)^{-1}\,(A-B)(I-sA)^{-1}$$

Let us note that

$$(I-sB)^{-1} = (I-sA)^{-1}(I+s(B-A)(I-sA)^{-1})^{-1}$$

and that for any operator K of trace class:

$$\text{Tr } K - \text{Tr } (I-sA)K(I-sA)^{-1}$$

Hence finally we can write:

$$\text{Tr.}(A(s) - B(s)) = \text{Tr. } (I+s(B-A)(I-sA)^{-1})^{-1} \quad (A-B)(I-sA)^{-2}$$

Let us observe now that because (A-B) is trace-class we can write the right side as:

$$\frac{d}{ds} \quad \text{Log Det } (I + C(s))$$

where

$$C(s) = s(A-B)(I-sA)^{-1}$$

and

$$\frac{d}{ds} \; C(s) = (A-B)(I-sA)^{-2}$$

But

$$I + C(s) = I + s(A-B)(I-sA)^{-1}$$

$$= (I-sB)(I-sA)^{-1}$$

and thus we have established (AI-2). The last step of course is the one that shows that the determinant in (AI-2) is well-defined! Next take

$$B = L^*$$
$$A = (-1)\,L(I-L)^{-1}$$

Then

$$B-A = L^* + L(I - L)^{-1}$$

But

$$(I-L)^{-1} = (I-M)$$

where $M$ is also Volterra, and Hilbert-Schmidt. Hence $LM$ is trace-class, and hence so is:

$$B-A = L^* + L - LM$$

Of course since $L$ is a Volterra operator, $(I-sA)$ and $(I-sB)$ have bounded inverses for all non-zero $s$. In particular then, using (AI-2), we obtain

$$\text{Log Det } (I-B)(I-A) = \int_0^1 \text{Tr}\,(A(s) - B(s))ds$$

In the present case:

$$A(s) - B(s) = (1/s)((I-sA)^{-1} - (I-sB)^{-1})$$

$$= (1/s)((I+sL(I-M))^{-1} - (I-sL^*)^{-1})$$

$$= (1/s)(I+sL\sum_0^{\infty}(-M)^k - I - \sum_1^{\infty} s^k L^{*k})$$

where the expansions are valid because $L$ is a Volterra operator, of Hilbert-Schmidt type. Again because of that

$$\text{Tr. } M^k = 0 \text{ for } k \geq 2$$

$$\text{Tr. } L^{*k} = 0 \text{ for } k \geq 2$$

$$\text{Tr. } LM = 0$$

so that we obtain:

$$\text{Tr. } (A(s) - B(s)) = (1/s)(\text{Tr. }(sL + sL^*))(-1)$$

and hence the result follows.

## APPENDIX II

### Krein Factorization Theorem

We shall consider only a version of the Krein factorization theorem [11] that is essential to our purposes. Thus let $r(s,t)$ be a continuous 'covariance' m-by-m matrix kernel, $0 \le s,t \le T$. The term 'covariance' means that the matrix:

$$\sum_{1}^{n} \sum_{1}^{n} a_i r(s_i;s_j)\bar{a}_j \qquad 0 \le s_i \le T$$

is non-negative definite, for any arbitrary choice of $\{s_i\}$, and n. We assume that $r(s,t)$ is real valued. Let $H(t)$ denote the $L_2$ space of m-by-one matrix functions on $[0,t]$. On $H(T)$ introduce the operator: $R(T)$ by:

$$R(T)f = g; \quad g(t) = \int_0^T r(t;s)\, f(s)ds, \quad 0 \le t \le T$$

mapping $H(T)$ into itself. Then we have:

Theorem (Krein):

$$(I + R(T))^{-1} = (I - \mathscr{L}^*)(I - \mathscr{L}) \qquad \text{(AII-}$$

where $\mathscr{L}$ is a Volterra operator:

$$\mathscr{L}f = g; \quad g(t) = \int_0^t L(t;s)f(s)ds \qquad \text{(AII-}$$

of Hilbert-Schmidt type.

Proof  Let $R(t)$ denote the operator mapping $H(t)$ into itself:

$$R(t)f = g; \quad g(s) = \int_0^t r(s;\sigma)\, f(\sigma)d\sigma, \quad 0 \leq s \leq t$$

Then $R(t)$ is non-negative definite by virtue of the covariance property, and since $r(\cdot;\cdot)$ is continuous, it is trace-class. Hence $(I + R(t))$ has a bounded inverse for each $t$. Let us take:

$$(I + R(t))^{-1} = (I - G(t)) \tag{AII-3}$$

Then $G(t)$ is an integral operator with kernel $G(t;s;\sigma)$:

$$G(t)f = g; \quad g(s) = \int_0^t G(t;s;\sigma)f(\sigma)d\sigma, \quad 0 \leq s \leq t$$

What is crucial however is that $G(t;s;\sigma)$ is continuous in the region $0 \leq s, \sigma \leq t \leq T$, and has continuous partial derivatives with respect to $t$. This follows from the Fredholm theory as noted in [11]. Now from (AII-3) it follows that

$$0 = R(t) - G(t) - R(t)G(t)$$

$$0 = R(t) - G(t) - G(t)R(t)$$

and hence in terms of corresponding kernels we have:

$$r(s;\sigma) = G(t;s;\sigma) + \int_0^t r(s;\tau)\, G(t;\tau;\sigma)\, d\tau, \quad 0 \leq s, \sigma \leq t \tag{AII-4}$$

We may now differentiate this partially with respect to $t$ and obtain:

$$(-1)H(t;s;\sigma) = r(s;t)G(t;t;\sigma) + \int_0^t r(s;\tau)H(t;\tau;\sigma)d\tau \tag{AII-5}$$

where

$$\frac{\partial}{\partial t} G(t;s;\sigma) = H(t;s;\sigma)$$

From (AII-4) putting $\sigma = t$ therein, we have:

$$r(s;t) = G(t;s;t) + \int_0^t r(s;\tau)\, G(t;\tau;t)d\tau$$

Multiplying both sides by $G(t;t;\sigma)$ we have:

$$r(s;t)G(t;t;\sigma) = G(t;s;t)G(t;t;\sigma) + \int_0^t r(s;\tau)G(t;\tau;t)G(t;t;\sigma)d\tau$$

while from (AII-5) the right side is also equal to

$$= (-1)H(t;s;\sigma) + \int_0^t r(s;\tau)(-1)H(t;\tau;\sigma)d\tau$$

By the non-singularity of $(I+R(t))$ this implies that for $0 \leq s, \sigma \leq t$:

$$H(t;s;\sigma) = (-1)G(t;s;t)G(t;t;\sigma) \qquad \text{(AII}$$

This is the central result we need. Now define the operator $\mathscr{L}$ on $H(T)$ by:

$$\mathscr{L}f = g; \quad g(t) = \int_0^t G(t;t;s)f(s)ds \qquad 0 \leq t \leq T$$

mapping $H(T)$ into itself. Note that $G(t;t;s)$ is continuous in the triangle $s \leq t \leq T$. Note that

$$[\mathscr{L}f, h] = \int_0^T [f(s), \int_s^T G(t;t;s)^* h(t)dt]$$

But because $G(t)$ is self-adjoint, we must have:

$$G(t;\sigma;s)^* = G(t;s;\sigma)$$

and hence with $s \leq t$:

$$G(t;t;s)^* = G(t;s;t)$$

Hence $\mathscr{L}^*$ is defined by:

$$\mathscr{L}^* h = g; \quad g(t) = \int_t^T G(s;t;s)\, h(s) ds$$

To show that this yields the factorization sought, it is enough to show that

$$G(T) = \mathscr{L} + \mathscr{L}^* - \mathscr{L}^* \mathscr{L} \qquad \text{(AII-7)}$$

or, that for $s < t$:

$$G(T;t;s) = G(t;t;s) - \int_t^T G(\tau;t;\tau)\, G(\tau;\tau;s) d\tau \qquad \text{(AII-8)}$$

But from (AII-6) we note that for $\tau \geq t \geq s$:

$$(-1) G(\tau;t;\tau)\, G(\tau;\tau;s) = \frac{\partial}{\partial \tau} G(\tau;t;s)$$

and hence the integral term in (AII-8) reduces to

$$G(T;t;s) - G(t;t;s)$$

thus verifying the equality (AII-7). The result for $t < s$ is clearly established similarly. Hence (AII-7) follows, and hence (AII-1). Note that the kernel for $\mathscr{L}$ does not depend on the final end-point T.

## REFERENCES


1. A. N. Kolmogorov: Foundations of the Theory of Probability, Chelsea, 1950.

2. J. L. Doob: Stochastic Processes, John Wiley and Sons, 1953.

3. L. I. Gikhman and A. V. Skorokhod: Introduction to the Theory of Random Processes, W. B. Saunders, 1969.

4. K. Parthasarathy: "Probability Measures on Metric Spaces", Academic Press, 1967.

5. L. Shepp: Radon-Nikodym Derivatives of Gaussian Measures: Annals of Mathematical Statistics, 1966.

6. H. McKean: Stochastic Integrals, Academic Press. 19

7. W. M. Wonham: "Random Differential Equations in Control Theory", in Probabilistic Methods in Applied Mathematics, Volume 2, Academic Press, 1970.

8. E. Nelson: "Dynamical Theories of Brownian Motion", Princeton University Press, 1967.

9. W. M. Wonham: "On a Matrix Riccati Equation of Stochastic Control", SIAM Journal on Control, Volume 6, No. 4, 1968.

10. M. Loeve: Probability Theory, Van Nostrand, 1954.

11. I. Gohberg and M. G. Krein: Volterra Operators, AMS Translations, 1970.

12. P. D. Joseph and J. T. Tou: "On Linear Control Theory", AIEE Transactions, Applications and Industry, Vol. 80, p. 193-196, 1961.

13. S. Karlin: "Mathematical Methods and Theory in Games, Programming and Economics", Vol. 2, Addison-Wesley, 1959.

14. H. Cramer: Methods of Mathematical Statistics, Princeton University Press, 1947.

REFERENCES (continued)

15. I. C. Gohberg and M. G. Krein: Introduction to the Theory of Linear Non-Self-Adjoint Operators, Vol. 18, Translations of Mathematical Monographs.

16. T. Kailath: An Innovations Approach to Least Squares Estimation, IEEE Transactions on Automatic Control, December 1968.

17. H. Kushner: "Stochastic Stability and Control", Academic Press, 1967.

18. L. Wong: "Stochastic Processes in Information and Dynamic Systems", McGraw-Hill, 1971.

19. E. McShane: 'Stochastic Equations and Stochastic Models', Holt, Rhinehart & Whinston, 1972.

20. E. B. Dynkin: Markov Process, Vol. I & II, Academic Press, 1965.

21. A. E. Bryson and Y. C. Ho: Applied Optimal Control, Ginn & Co., 1969.

22. R. S. Bucy and P. D. Joseph: 'Filtering for Stochastic Processes with Applications to Guidance', Interscience Publishers, 1968.

23. I. M. Gelfand and N. Ya. Vilenkin, Generalized Functions, Vol. 4, Academic Press, 1964.

24. J. Neveu: Mathematical Foundations of the Calculus of Probability, Holden Day, 1965.

25. R. Isaacs: Differential Games, John Wiley & Sons, 1967.

26. K. Ito: On Stochastic Differential Equations, Memoir, American Mathematical Society, 1951.

27. Y. A. Rozanov: Stationary Random Processes, Holden Day, 1967.

28. K. T. Woo: "Maximum Likelihood Estimation of Noisy Systems", Dissertation, UCLA Engineering, 1969.

29. W. H. Fleming: Optimal Continuous Parameter Stochastic Control, SIAM Review, Vol. 11, No. 4, October 1969.

REFERENCES (continued)


30. A. V. Skorokhod: Studies in the Theory of Random Processes, Addison-Wesley, 1965.

31. P. Levy: 'Processus Stochastiques et Mouvement Brownien', Gauthier-Villars, Paris, 1948.

32. N. Dunford and J. T. Schwartz: Linear Operators, Part II, Spectral Theory, Interscience Publishers, 1963.

33. Yu. V. Rozanov, Yu. A. Prokhrov: 'Probability Theory': Springer Verlag, 1969.

34. H. F. Karreman (Editor): 'Stochastic Optimization and Control', John Wiley and Sons, 1968.

35. A. V. Balakrishnan and L. W. Neustadt (Editors): 'Mathematical Theory of Control', Academic Press, 1967.

36. A. V. Balakrishnan: "Identification of Linear Dynamic Systems", Proceedings of the 3rd Annual South Eastern Symposium on System Theory, Atlanta, Georgia, April 1971.

37. A. V. Balakrishnan: "Stochastic Control: Function Space Approach", Journal of SIAM on Control, Vol. 10, No. 2, May 1972.

38. K. J. Astrom: Introduction to Stochastic Control Theory, Academic Press, 1970.

39. H. Kushner: Introduction to Stochastic Control Theory, Holt, Rinehart and Winston, 1971.

40. A. V. Balakrishnan: "Communication Theory", McGraw Hill Book Co., 1968.

41. A. V. Balakrishnan: 'Identification and Adaptive Control: An Application to Flight Control Systems', Journal of Optimization Theory and Applications, Vol. 9, No. 3, March 1972.

42. K. J. Aström and P. Eykhoff: Survey Paper: 'System Identification', Proceedings of the Second IFAC Symposium on Identification and Process Parameter Estimation, Prague, Czechoslovakia, 1970.

# SUPPLEMENTARY NOTES

## INTRODUCTION

Although we have dealt exclusively with continuous time models, any extensive practical use of the theories is feasible only with high speed digital computation - and hence the continuous time data must be 'sampled' and the Ito integrals that occur must be approximated by 'sums'. We discuss this aspect briefly in these Notes. We begin with the sampled or discrete version of recursive filtering and likelihood ratios, because of its independent interest.

## 1. DISCRETE (SAMPLED DATA) RECURSIVE FILTERING:

Because our continuous time theory extends (or specializes!) readily to the case where we have a time series instead of continuous time process, and because the approach is different from extant treatment of the topic in textbooks, we shall now consider the case where (omitting the sample point indication hereto):

$$y_n = C_n x_n + G_n W_n \qquad n \geq 1 \tag{1}$$

$$x_{n+1} = A_n x_n + F_n W_n \qquad n \geq 0, \quad x_0 = 0 \tag{2}$$

Here $W_n$ is a sequence of independent Gaussians with unit covariance matrix, and it is assumed that

$$G_n G_n^* > 0$$

for every n. Moreover for simplicity we shall confine ourselves to the case where signal and noise are independent:

$$F_n G_n^* = 0$$

It is desired to find:

$$\hat{x}_n = E(x_n / y_n, \ldots . y_1)$$

Imitating our treatment of the continuous case we now observe:

Step 1: Let

$$z_n^o = y_n - C_n A_{n-1} \hat{x}_{n-1} \tag{3}$$

Then $z_n^o$ is a sequence of zero mean independent Gaussians. To see this let

$$e_n = x_n - \hat{x}_n \quad ; \quad P_n = E((e_n e_n^*))$$

Then

$$z_n^o = C_n A_{n-1} e_{n-1} + C_n F_{n-1} W_{n-1} + G_n W_n$$

and each term herein is independent of the sigma-algebra generated by $y_{n-1}, \ldots y_1$, and hence also of $z_m^o$ for $m$ less than $n$.

Step 2: The sequence:

$$\hat{x}_n - A_{n-1} \hat{x}_{n-1} = z_n^s \tag{4}$$

where the superscript 's' stands for 'state' is a sequence of independent Gaussians. For, we have only to note that we can express $z_n^s$ as:

$$z_n^s = -(x_n - \hat{x}_n) + A_{n-1}(x_{n-1} - \hat{x}_{n-1}) + F_{n-1} W_{n-1}$$

$$= -e_n + A_{n-1} e_{n-1} + F_{n-1} W_{n-1}$$

and we note that $e_n$ is independent of the sigma algebra generated by $y_n, \ldots y_1$, and $W_n$ is of course independent of all variables with indices less than n, and is in addition independent of $x_n$. Hence it follows that $z_n^s$ is actually independent of $z_m^s$ for m less than n.

Step 3: The sigma-algebras $\mathscr{B}(y_n, \ldots y_1)$ and $\mathscr{B}(z_n^o, \ldots z_1^o)$ are equivalent. This is immediate from the fact that we can explicitly calculate the $y_n, \ldots y_1$, from the $z_n^o, \ldots z_1^o$ from the definition of the latter (by inverting a matrix which is the sum of the Identity matrix plus triangular matrix with zeros along the diagonal). Hence

$$z_n^s = K_n z_n^o$$

where $K_n$ satisfies:

$$E(z_n^s z_n^{o*}) = K_n E(z_n^o z_n^{o*})$$

Step 4: It only remains to calculate the matrices needed for finding $K_n$. We note that $F_{n-1} W_{n-1}$ is independent of $x_{n-1}$ and $y_{n-1}, \ldots y_1$ and hence also of $e_{n-1}$; (and this is the only place where the independence of signal and noise ($F_n G_n^* = 0$) is used). Hence

$$E(z_n^o z_n^{o*}) = C_n A_{n-1} P_{n-1} A_{n-1}^* C_n^*$$

$$+ C_n F_{n-1} F_{n-1}^* C_n^* + G_n G_n^* \qquad (5)$$

Note that this matrix is non-singular because of our assumption that

$G_n G_n^*$ is. Next

$$E(z_n^s z_n^{o*}) = E((-e_n + A_{n-1} e_{n-1} + F_{n-1} W_{n-1})(y_n - C_n A_{n-1} \hat{x}_{n-1})^*)$$

$$= E((A_{n-1} e_{n-1} + F_{n-1} W_{n-1})(C_n A_{n-1} e_{n-1} + C_n F_{n-1} W_{n-1} + G_n W_n)^*)$$

$$= A_{n-1} P_{n-1} A_{n-1}^* C_n^* + F_{n-1} F_{n-1}^* C_n^*$$

Hence

$$K_n = (A_{n-1} P_{n-1} A_{n-1}^* C_n^* + F_{n-1} F_{n-1}^* C_n^*)(C_n A_{n-1} P_{n-1} A_{n-1}^* C_n^*$$

$$+ C_n F_{n-1} F_{n-1}^* C_n^* + G_n G_n^*)^{-1}$$

And finally: we have the recursive equations characterizing $\hat{x}_n$:

$$\hat{x}_{n+1} = A_n \hat{x}_n + K_{n+1} z_{n+1}^o$$

$$z_{n+1}^o = y_{n+1} - C_{n+1} A_n \hat{x}_n$$

It only remains to obtain the difference equation characterizing $P_n$. For this we note that

$$P_n = E(x_n x_n^*) - E(\hat{x}_n \hat{x}_n^*)$$

where

$$E(x_n x_n^*) = A_{n-1} E(x_{n-1} x_{n-1}^*) A_{n-1}^* + F_{n-1} F_{n-1}^*$$

$$E(\hat{x}_n \hat{x}_n^*) = A_{n-1} E(\hat{x}_{n-1} \hat{x}_{n-1}^*) A_{n-1}^* + K_n (C_n A_{n-1} P_{n-1} A_{n-1}^*$$

$$+ C_n F_{n-1} F_{n-1}^*)$$

so that

$$P_{n+1} = A_n P_n A_n^* + F_n F_n^*$$

$$-(A_n P_n A_n^* C_{n+1}^* + F_n F_n^* C_{n+1}^*)$$

$$(C_{n+1} A_n P_n C_{n+1}^* + C_{n+1} F_n F_n^* C_{n+1}^* + G_{n+1} G_{n+1}^*)^{-1}$$

$$C_{n+1}(A_n P_n A_n^* + F_n F_n^*) \qquad (7)$$

$$P_0 = 0$$

## 2. LIKELIHOOD RATIOS: DISCRETE CASE

Under the same set up as in #1, let us examine the likelihood ratio based on N consecutive data samples. Thus let Y denote the vector of N samples $y_1, \ldots y_N$; let $p(Y)$ denote the corresponding probability density, and let $G(Y)$ density in the absence of signal.(taking $C_n$ to be zero).

Then we have

$$\text{Log } p(Y)/G(Y) = -\frac{1}{2}\left(\sum_1^N [D_n^{-1} z_n^o, z_n^o] - \sum_1^N [(G_n G_n^*)^{-1} y_n, y_n]\right.$$

$$-\frac{1}{2}\sum_1^N \text{Log (Det. } D_n/\text{Det. } G_n G_n^*) \qquad (8)$$

where

$$D_n = E(z_n^o z_n^{o*})$$

To see the similarity to the continuous case, let $\hat{Y}$ denote the vector composed of $C_1 A_0 \hat{x}_0$, $C_2 A_1 \hat{x}_1$, ...., $C_N A_{N-1} \hat{x}_{N-1}$ and $\mathscr{L}$ denote the matrix where

$$\hat{Y} = \mathscr{L} Y$$

Then $\mathscr{L}$ is a triangular matrix with zeros along the diagonal (so that in particular Tr. $\mathscr{L} = 0$). We then have:

Log $p(Y)/G(Y)$

$$= (-\frac{1}{2})\left\{[\mathscr{D}^{-1}(I-\mathscr{L})Y, (I-\mathscr{L})Y] - [\mathscr{G}^{-1}Y, Y]\right\} - \frac{1}{2}\text{ Tr. Log }\mathscr{D} + \frac{1}{2}\text{ Tr. Log }\mathscr{G}$$

where the matrix $\mathscr{D}$ is composed as

$$\mathscr{D} = \text{diagonal }(D_1, D_2, \ldots D_N)$$

and similarly

$$\mathscr{G} = \text{diagonal }(G_1, G_2, \ldots G_n)$$

The appearance of the third term containing the Trace should cause no surprise; we have already seen it in connection with the Radon-Nikodym derivative in Chapter 5, and we have seen how the Ito integral eliminates it. We shall explore this point a little further. Suppose we consider the system we studied for the Identification problem, Chapter 8. Let

$$x(t) = \int_0^t A\, x(s)\, ds + \int_0^t B\, u(s)\, ds + \int_0^t F\, dW(s) \tag{9}$$

$$y(s) = \int_0^t C\, x(s)\, ds + \int_0^t D\, u(s) + G\int_0^t dW(s) \tag{10}$$

where we have again suppressed the sample point '$\omega$'. We assume

$$F\, G^* = 0\,; \quad GG^* = I$$

Here $y(s)$ is the observation. In many practical situations however this model for the observation is not quite accurate. (One can argue generally in fact that a true Wiener process can never be observed; that any physical measuring apparatus must eventually cease to respond as the input frequencies increase beyond limit.) In fact what is observed is

$$v(t) = \int_0^t h(t-s)\, dy(s) \tag{11}$$

where

$$\int_0^\infty \| h(t) \|^2 \, dt < \infty$$

Further, if digital computer processing of the data is involved, as is the case inevitably if this level of processing sophistication is to be accepted, then what is observed are samples at periodic intervals; in other words:

$$v(n\Delta)$$

where $\Delta$ is sufficiently small time interval. We assume that $\Delta$ is small enough so that the measuring apparatus does not distort the 'signal s' $u(t)$, $x(t)$. A reasonable approximate model is to take

$$v(n\Delta) = \frac{y(\overline{n+1}\,\Delta) - y(n\Delta)}{\Delta}$$

$$v(n\Delta) \doteqdot C\, x(n\Delta) + D\, u(n\Delta) + G\, \zeta_n / \sqrt{\Delta}$$

where

$$\zeta_n = \frac{W(\overline{n+1}\,\Delta) - W(n\Delta)}{\sqrt{\Delta}}$$

so that,

$$E(\zeta_n \zeta_m^*) = 0 \quad m \neq n$$

$$= I \quad m = n.$$

Note that

$$E[\|v(n\Delta) - C\,x(n\Delta) - D\,u(n\Delta)\|^2] = (\mathrm{Tr.}\ GG^*)(1/\Delta)$$

and $\to \infty$ as $\Delta \to 0$.

[Another interpretation is that the observation is 'band-limited' to $[-\frac{1}{2\Delta}, \frac{1}{2\Delta}]$; (See [40] for more on 'band-limited' signals); and then sampled at periodic $\Delta$ intervals.] We can then apply the discrete theory if we assume $\Delta$ is small enough so that we can in addition use the representation:

$$x(\overline{n+1}\,\Delta) = e^{A\Delta} x(n\Delta) + e^{A\Delta} B\,u(n\Delta)\,\Delta + e^{A\Delta} F\xi(n\Delta)\sqrt{\Delta},$$

which arises by the approximation

$$\int_{n\Delta}^{(n+1)\Delta} e^{A(\overline{n+1}\,\Delta - s)} B\,u(s)ds \doteq e^{A\Delta} B\,u(n\Delta)\Delta$$

$$\int_{n\Delta}^{(n+1)\Delta} e^{A(\overline{n+1}\,\Delta - s)} F\,dW(s) \doteq e^{A\Delta} F(W(\overline{n+1}\,\Delta) - W(n\Delta))$$

so that $\{\xi(n\Delta)\}$ is a sequence of independent zero mean unit variance Gaussians. The approximation error is of the order of $\Delta$ assuming that the function $u(t)$ is bounded on $[0, \infty)$, as a ready calculation shows. We are then able to apply the discrete version theory by setting:

$$A_n = e^{A\Delta}$$

$$F_n = F\sqrt{\Delta}$$

$$G_n = (1/\sqrt{\Delta})G$$

$$y_n = v(n\Delta) - m(n\Delta)$$

where

$$m(n\Delta) = C\,S(n\Delta) + D\,u(n\Delta)$$

$$S(\overline{n+1}\,\Delta) = S(n\Delta) + e^{A\Delta} B\,u(n\Delta)\Delta\ ;\ S(0) = 0$$

We thus obtain for the (log) likelihood ratio based on N consecutive samples, $y_1, \ldots y_N$:

$$-\frac{1}{2}\left\{\sum_1^N D_n^{-1}[z_n^o, z_n^o] - \sum_1^N [y_n, y_n]\right\} - \frac{1}{2}\sum_1^N \text{Log Det}(D_n) \tag{12}$$

Let us note that

$$\Delta D_n = (C\,e^{A\Delta} P_{n-1}\, e^{A^*\Delta} C^*\Delta + C\,F\,F^*C^*\,\Delta^2 + I)$$

For $\Delta$ sufficiently small we note the approximation:

$$\text{Log Det}(\Delta D_n) = \text{Trace Log}\ \Delta D_n$$

$$\doteq \Delta\ \text{Trace}\ CP_{n-1}C^*$$

with the error of the order of $\Delta^2$. Also

$$[D_n^{-1} z_n^o, z_n^o] \doteq \Delta[z_n^o, z_n^o]$$

$$z_n^o \doteq y_n - C\hat{x}_n$$

$$K_n \doteq P_{n-1}C^*\,\Delta$$

If we go along with the interpretation that $v(n\Delta)$ are samples after 'band-limiting' to $(-\frac{1}{2\Delta}, \frac{1}{2\Delta})$, then $v(t)$ denoting the band-limited observation, the discrete sums in the (log) likelihood ratio approximate the integrals:

$$(-\tfrac{1}{2}) \int_0^T (\|\tilde{v}(t) - \hat{\tilde{v}}(t)\|^2 - \|v(t)\|^2)dt \; - \tfrac{1}{2}\int_0^T \text{Tr. } CP(t)C^* \, dt \tag{13}$$

where

$$\tilde{v}(t) = v(t) - E(v(t))$$

$$\hat{\tilde{v}}(\cdot) = \mathscr{L}\, \tilde{v}(\cdot)$$

the operator $\mathscr{L}$ being given by:

$$\mathscr{L} f = g; \; g(t) = C\, \phi(t) \int_0^t \phi(s)^{-1}\, P(s)\, C^*\, f(s)\, ds$$

where

$$\dot{\phi}(t) = (A - P(t)C^*C)\, \phi(t)$$

There is a difficultyin this interpretation in that the band-limiting operation is not physically realizable, strictly speaking; one must accept a 'time-delay', which is theoretically 'infinite'. Perhaps a better view is to consider (12) as the correct 'discrete' approximation to the Ito integral occurring in the formula for the likelihood ratio in (8.4). The appearance of the trace term is the 'penalty' one has to pay in this kind of approximation - even though it is the only natural one in most situations.

# Lecture Notes in Economics and Mathematical Systems

(Vol. 1–15: Lecture Notes in Operations Research and Mathematical Economics, Vol. 16–59: Lecture Notes in Operations Research and Mathematical Systems)

Vol. 1: H. Bühlmann, H. Loeffel, E. Nievergelt, Einführung in die Theorie und Praxis der Entscheidung bei Unsicherheit. 2. Auflage, IV, 125 Seiten 4°. 1969. DM 16,–

Vol. 2: U. N. Bhat, A Study of the Queueing Systems M/G/1 and GI/M/1. VIII, 78 pages. 4°. 1968. DM 16,–

Vol. 3: A. Strauss, An Introduction to Optimal Control Theory. VI, 153 pages. 4°. 1968. DM 16,–

Vol. 4: Branch and Bound: Eine Einführung. 2., geänderte Auflage. Herausgegeben von F. Weinberg. VII, 174 Seiten. 4°. 1972. DM 18,–

Vol. 5: Hyvärinen, Information Theory for Systems Engineers. VIII, 205 pages. 4°. 1968. DM 16,–

Vol. 6: H. P. Künzi, O. Müller, E. Nievergelt, Einführungskursus in die dynamische Programmierung. IV, 103 Seiten. 4°. 1968. DM 16,–

Vol. 7: W. Popp, Einführung in die Theorie der Lagerhaltung. VI, 173 Seiten. 4°. 1968. DM 16,–

Vol. 8: J. Teghem, J. Loris-Teghem, J. P. Lambotte, Modèles d'Attente M/G/1 et GI/M/1 à Arrivées et Services en Groupes. IV, 53 pages. 4°. 1969. DM 16,–

Vol. 9: E. Schultze, Einführung in die mathematischen Grundlagen der Informationstheorie. VI, 116 Seiten. 4°. 1969. DM 16,–

Vol. 10: D. Hochstädter, Stochastische Lagerhaltungsmodelle. VI, 269 Seiten. 4°. 1969. DM 18,–

Vol. 11/12: Mathematical Systems Theory and Economics. Edited by H. W. Kuhn and G. P. Szegö. VIII, IV, 486 pages. 4°. 1969. DM 34,–

Vol. 13: Heuristische Planungsmethoden. Herausgegeben von F. Weinberg und C. A. Zehnder. II, 93 Seiten. 4°. 1969. DM 16,–

Vol. 14: Computing Methods in Optimization Problems. Edited by A. V. Balakrishnan. V, 191 pages. 4°. 1969. DM 16,–

Vol. 15: Economic Models, Estimation and Risk Programming: Essays in Honor of Gerhard Tintner. Edited by K. A. Fox, G. V. L. Narasimham and J. K. Sengupta. VIII, 461 pages. 4°. 1969. DM 24,–

Vol. 16: H. P. Künzi und W. Oettli, Nichtlineare Optimierung: Neuere Verfahren, Bibliographie. IV, 180 Seiten. 4°. 1969. DM 16,–

Vol. 17: H. Bauer und K. Neumann, Berechnung optimaler Steuerungen, Maximumprinzip und dynamische Optimierung. VIII, 188 Seiten. 4°. 1969. DM 16,–

Vol. 18: M. Wolff, Optimale Instandhaltungspolitiken in einfachen Systemen. V, 143 Seiten. 4°. 1970. DM 16,–

Vol. 19: L. Hyvärinen, Mathematical Modeling for Industrial Processes. VI, 122 pages. 4°. 1970. DM 16,–

Vol. 20: G. Uebe, Optimale Fahrpläne. IX, 161 Seiten. 4°. 1970. DM 16,–

Vol. 21: Th. Liebling, Graphentheorie in Planungs- und Tourenproblemen am Beispiel des städtischen Straßendienstes. IX, 118 Seiten. 4°. 1970. DM 16,–

Vol. 22: W. Eichhorn, Theorie der homogenen Produktionsfunktion. VIII, 119 Seiten. 4°. 1970. DM 16,–

Vol. 23: A. Ghosal, Some Aspects of Queueing and Storage Systems. IV, 93 pages. 4°. 1970. DM 16,–

Vol. 24: Feichtinger, Lernprozesse in stochastischen Automaten. V, 66 Seiten. 4°. 1970. DM 16,–

Vol. 25: R. Henn und O. Opitz, Konsum- und Produktionstheorie. I. II, 124 Seiten. 4°. 1970. DM 16,–

Vol. 26: D. Hochstädter und G. Uebe, Ökonometrische Methoden. XII, 250 Seiten. 4°. 1970. DM 18,–

Vol. 27: I. H. Mufti, Computational Methods in Optimal Control Problems. IV, 45 pages. 4°. 1970. DM 16,–

Vol. 28: Theoretical Approaches to Non-Numerical Problem Solving. Edited by R. B. Banerji and M. D. Mesarovic. VI, 466 pages. 4°. 1970. DM 24,–

Vol. 29: S. E. Elmaghraby, Some Network Models in Management Science. III, 177 pages. 4°. 1970. DM 16,–

Vol. 30: H. Noltemeier, Sensitivitätsanalyse bei diskreten linearen Optimierungsproblemen. VI, 102 Seiten. 4°. 1970. DM 16,–

Vol. 31: M. Kühlmeyer, Die nichtzentrale t-Verteilung. II, 106 Seiten. 4°. 1970. DM 16,–

Vol. 32: F. Bartholomes und G. Hotz, Homomorphismen und Reduktionen linearer Sprachen. XII, 143 Seiten. 4°. 1970. DM 16,–

Vol. 33: K. Hinderer, Foundations of Non-stationary Dynamic Programming with Discrete Time Parameter. VI, 160 pages. 4°. 1970. DM 16,–

Vol. 34: H. Störmer, Semi-Markoff-Prozesse mit endlich vielen Zuständen. Theorie und Anwendungen. VII, 128 Seiten. 4°. 1970. DM 16,–

Vol. 35: F. Ferschl, Markovketten. VI, 168 Seiten. 4°. 1970. DM 16,–

Vol. 36: M. P. J. Magill, On a General Economic Theory of Motion. VI, 95 pages. 4°. 1970. DM 16,–

Vol. 37: H. Müller-Merbach, On Round-Off Errors in Linear Programming. VI, 48 pages. 4°. 1970. DM 16,–

Vol. 38: Statistische Methoden I, herausgegeben von E. Walter. VIII, 338 Seiten. 4°. 1970. DM 22,–

Vol. 39: Statistische Methoden II, herausgegeben von E. Walter. IV, 155 Seiten. 4°. 1970. DM 16,–

Vol. 40: H. Drygas, The Coordinate-Free Approach to Gauss-Markov Estimation. VIII, 113 pages. 4°. 1970. DM 16,–

Vol. 41: U. Ueing, Zwei Lösungsmethoden für nichtkonvexe Programmierungsprobleme. VI, 92 Seiten. 4°. 1971. DM 16,–

Vol. 42: A. V. Balakrishnan, Introduction to Optimization Theory in a Hilbert Space. IV, 153 pages. 4°. 1971. DM 16,–

Vol. 43: J. A. Morales, Bayesian Full Information Structural Analysis. VI, 154 pages. 4°. 1971. DM 16,–

Vol. 44: G. Feichtinger, Stochastische Modelle demographischer Prozesse. XIII, 404 Seiten. 4°. 1971. DM 28,–

Vol. 45: K. Wendler, Hauptaustauschschritte (Principal Pivoting). II, 64 Seiten. 4°. 1971. DM 16,–

Vol. 46: C. Boucher, Leçons sur la théorie des automates mathématiques. VIII, 193 pages. 4°. 1971. DM 18,–

Vol. 47: H. A. Nour Eldin, Optimierung linearer Regelsysteme mit quadratischer Zielfunktion. VIII, 163 Seiten. 4°. 1971. DM 16,–

Vol. 48: M. Constam, Fortran für Anfänger. VI, 143 Seiten. 4°. 1971. DM 16,–

Vol. 49: Ch. Schneeweiß, Regelungstechnische stochastische Optimierungsverfahren. XI, 254 Seiten. 4°. 1971. DM 22,–

Vol. 50: Unternehmensforschung Heute – Übersichtsvorträge der Züricher Tagung von SVOR und DGU, September 1970. Herausgegeben von M. Beckmann. VI, 133 Seiten. 4°. 1971. DM 16,–

Vol. 51: Digitale Simulation. Herausgegeben von K. Bauknecht und W. Nef. IV, 207 Seiten. 4°. 1971. DM 18,–

Vol. 52: Invariant Imbedding. Proceedings of the Summer Workshop on Invariant Imbedding Held at the University of Southern California, June – August 1970. Edited by R. E. Bellman and E. D. Denman. IV, 148 pages. 4°. 1971. DM 16,–

Vol. 53: J. Rosenmüller, Kooperative Spiele und Märkte. IV, 152 Seiten. 4°. 1971. DM 16,–

Vol. 54: C. C. von Weizsäcker, Steady State Capital Theory. III, 102 pages. 4°. 1971. DM 16,–

Vol. 55: P. A. V. B. Swamy, Statistical Inference in Random Coefficient Regression Models. VIII, 209 pages. 4°. 1971. DM 20,–

Vol. 56: Mohamed A. El-Hodiri, Constrained Extrema. Introduction to the Differentiable Case with Economic Applications. III, 130 pages. 4°. 1971. DM 16,–

Vol. 57: E. Freund, Zeitvariable Mehrgrößensysteme. VII, 160 Seiten. 4°. 1971. DM 18,–

Vol. 58: P. B. Hagelschuer, Theorie der linearen Dekomposition. VII, 191 Seiten. 4°. 1971. DM 18,–

Vol. 59: J. A. Hanson, Growth in Open Economics. IV, 127 pages. 4°. 1971. DM 16,–

Vol. 60: H. Hauptmann, Schätz- und Kontrolltheorie in stetigen dynamischen Wirtschaftsmodellen. V, 104 Seiten. 4°. 1971. DM 16,–

Vol. 61: K. H. F. Meyer, Wartesysteme mit variabler Bearbeitungsrate. VII, 314 Seiten. 4°. 1971. DM 24,–

Vol. 62: W. Krelle u. G. Gabisch unter Mitarbeit von J. Burgermeister, Wachstumstheorie. VII, 223 Seiten. 4°. 1972. DM 20,–

Vol. 63: J. Kohlas, Monte Carlo Simulation im Operations Research. VI, 162 Seiten. 4°. 1972. DM 16,–

Vol. 64: P. Gessner u. K. Spremann, Optimierung in Funktionenräumen. IV, 120 Seiten. 4°. 1972. DM 16,–

Vol. 65: W. Everling, Exercises in Computer Systems Analysis. VIII, 184 pages. 4°. 1972. DM 18,–

Vol. 66: F. Bauer, P. Garabedian and D. Korn, Supercritical Wing Sections. V, 211 pages. 4°. 1972. DM 20,–

Vol. 67: I. V. Girsanov, Lectures on Mathematical Theory of Extremum Problems. V, 136 pages. 4°. 1972. DM 16,–

Vol. 68: J. Loeckx, Computability and Decidability. An Introduction for Students of Computer Science. VI, 76 pages. 4°. 1972. DM 16,–